T0299722

Analysis of Questionnaire Data with R

Bruno Falissard

CRC Press
Taylor & Francis Group
Boca Raton   London   New York

CRC Press is an imprint of the
Taylor & Francis Group, an **informa** business

A CHAPMAN & HALL BOOK

CRC Press
Taylor & Francis Group
6000 Broken Sound Parkway NW, Suite 300
Boca Raton, FL 33487-2742

© 2012 by Taylor & Francis Group, LLC
CRC Press is an imprint of Taylor & Francis Group, an Informa business

No claim to original U.S. Government works

Version Date: 20110816

International Standard Book Number: 978-1-4398-1766-7 (Hardback)

Visit the Taylor & Francis Web site at
http://www.taylorandfrancis.com

and the CRC Press Web site at
http://www.crcpress.com

Contents

Preface

Most epidemiologists, sociologists, market research professionals and economists are now regularly dealing with quantitative data obtained from questionnaires. Sometimes these data are analyzed by a professional statistician, but most often it is the specialist in the particular field of study who does the job. In all situations, there is an absolute need to master statistics, a discipline often considered unpalatable because it lies at the intersection of two somewhat impenetrable disciplines—mathematics and data processing.

There is obviously a misunderstanding on this subject. If theoretical statistics do rely on mathematics, the practice of statistics does not, or in all events very little. Statistical practice is in fact like the process of translation, a process which, basically, proceeds from a hypothesis or a question formulated by a researcher to a series of variables linked by means of a statistical tool. As with written material, there is almost always a loss or a difference between the meaning of the original text and the meaning of the translated text, and many versions can generally be suggested, each with their advantages and disadvantages.

The objective of this book is to learn how to "translate" certain classic research questions into statistical formulations. As indicated in the title of the book, the syntax of these statistical formulations is based on the well-known R language. R has been chosen because it is accessed freely, and widely used, and above all because of the simplicity and the power of its structure: R helps to think statistics clearly. But syntax is not the main point; understanding the semantics is undoubtedly our real challenge. The semantics will emerge progressively from examples and experience, and only very occasionally from mathematical considerations.

Statistics are nothing without data. The part devoted to the analysis and the interpretation of examples is therefore important. These examples are all derived from the same genuinely implemented study (Mental Health in Prison study, 2006) with hundreds of variables measured on hundreds of subjects. This is an important particularity of questionnaire data: the number of variables and the number of subjects are both generally very large and, curiously, this has many practical consequences. The Mental Health in Prison study data set is available on the companion website to the book (http://bruno. falissard.pagesperso-orange.fr/AQR/index.html), with all the R syntaxes, so that it is easy for the reader to replicate and develop the analyses.

In real life, data sets contain errors, inconsistencies between answers, missing data, and so forth. Sometimes, the interpretation of a result is not clear. Sometimes, there is no statistical tool really suited to the question that is nagging you. More often, the statistical tool that you want to use is not formally appropriate, and it is difficult to assess to what extent this slight inadequacy affects the interpretation of results. These are also certain common situations that will be tackled in the following pages.

I hope that researchers and Master's or PhD students will find material here that will help them to gain some insight into their data. This is the ultimate achievement for a statistician.

Acknowledgments

My thanks to Florian Naudet and Christophe Lalanne for their helpful comments, and to Angela Verdier for her invaluable help with the English wording. They dealt with a previous version of the manuscript; therefore, the author is responsible for all that concerns the present one.

I am indebted to Anne Duburcq and Francis Fagnani for their fruitful collaboration in the Mental Health Prison study.

Finally, I especially thank my wife, Magali, son, Louis, and daughters Josephine and Violette, for their love and support.

1

Introduction

Statistics are used every day in very different areas, from astrophysicists observing billions of stars to biologists working on small samples of knock-out mice. This book deals with questionnaire data, that is, with responses that human beings have given to a series of standardized closed questions.

This brief introduction comprises four sections. Section 1.1 discusses a fundamental aspect of questionnaire data analysis while Section 1.2 focuses on basic technical and statistical aspects that will be developed in the remainder of the book. Section 1.3 presents the actual study that will be used as an illustration throughout the book: the MHP (Mental Health in Prison) study. Finally, Section 1.4 provides a very elementary "introduction to R."

1.1 About Questionnaires

Questionnaires collect human material, most often subjective; and while statistics can help interpret this material, to synthesise and to communicate it statistics alone will never be sufficient. Clearly, a sensible interpretation will emerge only if the researcher is familiar with the topic of the survey. Less obviously, this interpretation must also take into account the explicit and more often implicit traditions of the academic field of the researcher. Indeed, questionnaire surveys are implemented in disciplines as varied as market research, sociology, psychometrics, epidemiology, clinical research, demography, economics, political science, educational science, etc. And in all these academic specialties there are underlying theories concerning what happens when a human being answers questions: Is the investigator interested in the subject's opinions or only in the actual information contained in the answer? Concerning opinion, does it "really" exist? If it does, what is the definition of an "opinion"? If not, what is its epistemological status?

These problems are fascinating. They have led to a huge amount of work, not always consensual, and far beyond the scope of this book (which focuses essentially on statistical considerations). This is not to say that this corpus of research can be ignored, and we encourage readers who are less familiar with these issues to read basic references from their own field.

1.2 Principles of Analysis

1.2.1 Overviews

Several basic steps are involved quasi-systematically in the analysis of a dataset:

1. Data management
2. Description of variable distribution
3. Estimation of the strength of relationships between pairs of variables
4. Statistical tests of hypothesis
5. Statistical modelling

Data management concerns the import and the manipulation of datasets, as well as the verification and manipulation of variables. Questionnaire surveys often involve a large number of variables observed in a substantial sample of subjects so that data management is crucial.

The description of variable distribution is classically obtained through the estimation of means, variances, smallest and largest values, and graphical representations such as histograms or boxplots. Before the era of computers, statisticians had to perform all computations "by hand." This meant that they had to read and use all observations several times, and this gave them appreciable insight into variable distributions. This was precious at the time of interpretation. These days we can proceed much faster in our analyses, but we miss this comprehensive contact with the data. A systematic and meticulous look at the summary statistics of all variables is therefore necessary.

Many studies are performed because the investigators are looking for an association between two variables, for example, between a disease and a risk factor, or between the levels of consumption of two substances. Some parameters (e.g., relative risk and correlation coefficients) are designed to quantify the strength of association between two variables. Statistical tests of a hypothesis are designed to determine whether this association, or a stronger one, could have been observed solely by chance. Statistical tests are widely used and easy to implement in practice, but are, nevertheless, conceptually more complex than is generally assumed.

Finally, investigators are often interested in disentangling the pattern of associations that exists between a response (an outcome) on the one hand and a series of potential predictors on the other. Statistical models such as multiple linear regression and multiple logistic regression are dedicated to this. The design of a good model requires some experience, and in this area some points are inevitably raised: choice or the coding of predictors, determination of the relative importance of each predictor, potential interest of interaction terms, etc. One specificity of questionnaire data is that there being an outcome or a predictor is sometimes not very clear. In particular, a given variable can be a predictor of a second variable and an outcome of a third. Structural equation modelling can be useful in this situation.

We tackle all these points in the forthcoming pages, each time focusing on the specific features of questionnaire data.

1.2.2 Specific Aspects of Questionnaire Data Analysis

As mentioned above, a particularity of questionnaire data is that they are fundamentally open to interpretation. This has a statistical consequence: The meaning that can be associated with the response given to a particular item is most often clarified by the correlations that exist between this response and responses given to other questions. In a questionnaire survey, meaning and interpretations come to light progressively, each analysis providing an additional touch to a forthcoming global picture. This perhaps explains why exploratory analyses of correlations between variables are more often implemented here than in other areas of research. Exploratory data analysis will therefore occupy an important place in the remainder of the book.

Another feature of questionnaire data is that missing answers can be meaningful. Perhaps a subject did not understand the question, did not want to answer, or was not concerned by it. Sometimes, a non-response can even give an indication as to the value of the missing answers. For example, an absence of response to a question related to a high level of alcohol consumption is perhaps more likely to be observed in alcohol abusers. This particularity obviously has important statistical implications and a full section (6.6) will be devoted to it.

The question of measurement error, a cornerstone of the scientific method, raises specific issues in questionnaire surveys. The notion of reproducibility can be problematic. Psychometrics has developed interesting concepts that can help in determining the reliability of answers obtained to scales or questionnaires data. A Sections (7.5 and 7.6) are also devoted to this problem in the following pages.

1.3 The Mental Health in Prison (MHP) Study

In 2002, the French ministries of health and justice decided to determine the prevalence and risk factors of mental disorders in French prisons through an epidemiological study: the Mental Health in Prison (MHP) study. The main results and methodological aspects of the MHP study are presented in (Falissard et al. 2006). They are presented here briefly.

In 2004, there were in France approximately 65,000 individuals detained in three types of prison: The *Maisons d'arrêt* are intended for remand prisoners and/or for prisoners with short sentences; the *Maisons centrales* are intended for prisoners with long sentences, entailing maximum levels of security; and the *Centres de détention* are intended for those with intermediate sentences.

Prisoners were chosen at random using a two-stage stratified random sampling strategy: twenty prisons were first selected at random from the list of all French metropolitan prisons for men with stratification on the type of prison (*Maisons centrales* and *Centres de detention* were over-represented); second, prisoners were chosen at random in each of these 20 prisons until 800 prisoners were enrolled. At the time, 57% of the prisoners were available and agreed to the interview so that a sample of 1,402 prisoners was contacted between September 2003 and July 2004, producing a total of 799 interviews.

Each prisoner was interviewed for approximately two hours by two clinicians (clinical psychologist or psychiatrist), both of whom were present during the entire interview. At least one of the clinicians had to be a qualified psychiatrist; he (she) will be referred to as the "senior" member of the team. The interviews began with the collection of the signed informed consent of the prisoner. Psychiatric diagnoses were then recorded according to a semi-structured procedure: One of the clinicians used a structured clinical interview—the "M.I.N.I." (Mini-International Neuropsychiatric Interview) (Sheehan et al. 1995); the second, more experienced, completed the procedure with an open clinical interview of about 20 minutes, intended to be more clinically relevant. The interview continued with the completion of various socio-demographic questions, including personal, family, and judicial history. It terminated with the Temperament and Character Inventory (TCI, clinician version (Cloninger 2000)).

The main data files that will be used in the remainder of this book include the following:

"mhp.ex": This includes only nine variables and is used for elementary examples

"mhp.ex2": A slightly modified version of mhp.ex

"mhp.mod": The main dataset, it contains several hundred variables

"mhp.ira": This contains variables for the estimation of inter-rater agreement

"mhp.j" and "mhp.s": Two subsets obtained from the "junior" and "senior" interviewers (these two datasets are used in Chapter 9, which involves data manipulations)

1.4 If You Are a Complete R Beginner

1.4.1 First Steps

R is a language for statistical computing and not software that provides statistical analyses by simply clicking on menus and boxes. At first sight, R is not really user-friendly, and the few hours needed to master the commands required to engage in the analysis of a dataset may be daunting or dissuasive. These few hours are, however, clearly a good investment. The R language,

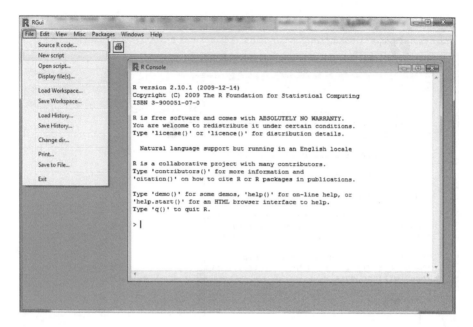

FIGURE 1.1
Screen obtained once R is launched. The "R Console" is designed to receive the commands. In return, it will provide the results. It can be useful to gather all the commands in a window set aside for that purpose. The instruction "New script" should be selected from the "File" menu for this purpose.

which is somewhat strict, indeed helps the user be rational and reasoned in his (her) statistical approach. In the medium and long term, the benefit to be drawn is considerable, especially if the user has no or only minimal training in mathematics and computer science. Another particularity of R is that the results are not simply presented in pages but are also stored in "objects". These objects have their own names and can be used as data in a new series of analyses. This can appear disconcerting and unnecessarily complex at first glance, but it turns out, in fact, to be a powerful property.

The R software can be downloaded and installed from http://cran.r-project.org/. Once R is launched, a screen corresponding to Figure 1.1 appears.

The main window, called the "R Console," receives the commands and returns the results. This is important in terms of quality control: Results are presented with the command line that generated them. Traceability is thus guaranteed. It is also easy to replicate the analyses several months or years later.

It can be useful to open a script window that will gather all the commands. This can be done by selecting the instruction "New script" in the "File" menu (Figure 1.1, then Figure 1.2).

A line of command ❶ can be written in this window. It is submitted to the "R Console" window by clicking in ❷. The result is obtained in ❸.

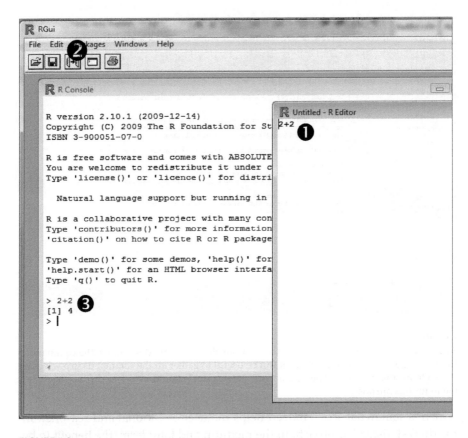

FIGURE 1.2
The window entitled "R Editor" is a script window that can gather instructions.

The most basic R instruction is the "<-" assignment. It can be used to store
a result into a variable:

```
       ❶
> x <- 10
> x❷
[1] 10❸
> y <- 20
> x + y❹
[1] 30
```

More precisely, if one wants to assign the number "10" to a variable that will
be labelled "x", then the instruction "<" (lower key) followed by "-" (dash key)
can be used ❶. If the name of the variable is written in a new command line
❷, its content is displayed in ❸. If "x" and "y" are two variables, the instruction "x + y" computes the sum of x and y ❹.

1.4.2 Functions from Optional Packages

Not all R functions are available in the standard version of the software. Many "packages" are available and must be installed separately. In practice, during an R session, select the option "install package(s)..." from the "Packages" menu. R then proposes selection of a mirror of the CRAN website close to your location and displays the list of all available packages among which the appropriate one can be selected. This manipulation requires an Internet connection but needs to be done only once (the package selected is stored with the programme). When a new R session is opened, each additional package must be installed using the function library()—for instance, library(prettyR) to install the "prettyR" package. All these packages are designed and programmed by fallible statisticians, which necessarily raises the question of the reliability of the functions they provide. The most efficient antidote against software bugs is likely to be intensive testing. Older packages have generally been used for a long time, and because the slightest surprising behaviour observed in a computation is in general presented on the R forum, with time a certain level of protection against the most salient errors develops. The packages used in this book include binom, boot, car, confint, Design, Epi, epicalc, gplots, Hmisc, lme4, ltm, MASS, mice, mgcv, nnet, plotrix, prettyR, psy, psych, relaimpo, RODBC, survey, and sem. To have these libraries loaded automatically when R is launched, you can add the following code to the "R_HOME\etc\Rprofile.site" file:

```
local({old <- getOption(defaultPackages")
options(defaultPackages = c(old, "binom", "boot", "car",
   "confint", "Design", "Epi", "epicalc", "gplots", "Hmisc",
   "lme4", "ltm", "MASS", "mice", "mgcv", "nnet", "plotrix",
   "prettyR", "psy", "psych", "relaimpo", "RODBC", "survey",
   "sem"))})
```

1.4.3 When Assistance Is Needed

The instruction "?" can be used when help is needed concerning a given function. For instance, information about the function hist(), which produces histograms, can be obtained from:

```
> ?hist
```

Examples are systematically provided and they are often particularly useful.

1.4.4 Importing a Dataset

Importing a dataset can be tricky, depending on the platform used (e.g., Windows PC, Macintosh, Linux, or Unix) and the origin of the file that contains the dataset: Is it from another statistical software program, from a spreadsheet application such as Microsoft Excel, or, more simply, is it in text format? Chapter 9, "Introduction to Data Manipulation with R," deals with these questions in more detail. A simple option is to open the file containing the dataset using a spreadsheet application; each column should correspond to a variable while the first row should contain the variable names. Then the "save as" command with the "csv" type option can be selected to save the dataset in a new "comma separated values" (csv) text format file. Finally, R can be opened, and the function read.csv() is used to import the "csv" text format file.

An error during the import process can have very serious consequences and, unfortunately, such errors are not rare. It is thus absolutely necessary to verify immediately that no major problem has occurred during the import process. The function str() is appropriate for this; it displays the data file content in compact form.

Let us now see this in an example. We intentionally use a very small data file (only nine variables). Once R has been launched, type the following in the R-console window or, better yet, in a script window:

❶   ❷   ❸   ❹
mhp.ex <- read.csv("C:/Users/Bruno/Book AQR/Data MHP/mhpex.csv")

The data file name and the directory path is in ❹ (Windows PC platform here. Note that the separators are "/" and not "\").

The function read.csv❸ imports the data file mhpex.csv. In countries where a comma is used as the decimal point, the function read.csv2 should be used instead of read.csv.

As seen above, the "<-" ❷ puts the data obtained by read.csv in an object that we have decided to call mhp.ex❶. From now on, this object is the data file on which we are going to work. In the terminology of R, it is called a "data frame."

1.4.5 More about the R Language

For readers not familiar with the R syntax and who want to learn more about it, it is possible at this stage to refer to Chapter 9 "Introduction to Data Manipulation with R."

2

Description of Responses

Mean, median, standard deviation, histogram, boxplot: Most of these parameters and graphics constitute the "basics of statistics." Their widespread use arises from the fact that with a few numbers and visual representations, they can simply and fairly comprehensively show how a variable is distributed, even if this involves thousands or millions of observations. Because they are the foundations, these analyses must be performed carefully, taking ample time to note and interpret each result. Numbers and percentages must be systematically set against the *a priori* representation that the statistician has of the data. When there is a discrepancy, it is essential to understand why and to check that there is no obvious error.

In this chapter we look first at the specific advantages of some of the most classic "summary statistics" (means, medians, standard deviations, minimum, maximum, quartiles, number of missing data, etc.). Then we concentrate on graphical representations such as histograms, boxplots, barplots, and pie charts.

2.1 Description Using Summary Statistics

In a few words: Even if the *mean* is a well-known parameter, it is not so easy to give it an intuitive meaning. From a geometrical point of view, it corresponds to the centre of gravity of the variable values. The median has a more straightforward meaning. It separates the higher half of the variable values. When the variable distribution is symmetrical, the mean is equal to the median.

While *mean* and *median* are two parameters that give an idea of the overall magnitude of the variable values, other parameters are intended to represent to what extent these values are distant from each other. Among these so-called "dispersion parameters," the standard deviation is likely the most familiar. The standard deviation has useful mathematical and even physical properties. Unfortunately, it has no intuitive meaning. As a rule of thumb, if the variable distribution is normal (i.e., if it has a bell curve shape), about two-thirds of the observations are between the mean plus/minus one standard deviation. The quartiles can also help assess the degree of dispersion of a variable. The lower quartile cuts off the lowest 25% of the variable values, while the upper quartile cuts off the highest 25%. Therefore, the 50% of the observations closest to the median lie between the lower and upper quartiles. This inter-quartile interval is favoured by many statisticians because it has an intuitive meaning and a simple formal definition.

The minimum and maximum for each variable are useful, in particular because they are efficient in detecting impossible or unusual observations (e.g., age of 490 years, which might correspond in fact to 49, with a 0 added by a slip on a computer key).

The number of missing values is also a parameter that is unavoidable: a variable with a high percentage of missing values must be used cautiously in sophisticated models, and the reason why so many missing values occurred must also be determined.

It is important to devote a number of hours to taking a critical look at the summary statistics for variables in a study. It is always a profitable investment.

In Practice: Taking a quick glance at the content of the data frame mhp.ex, the function str() is essential. The instruction str(mhp.ex) generates the following result:

```
> str(mhp.ex)
'data.frame': 799❶ obs. of 9❷ variables:
$ age❸      : int❹ 32 50 51 48 22 33 25 53 43 44❺ ...
$ prof      : Factor w/ 7 levels "farmer", "craftsman", ..:
              7 NA 4 6 NA 6 7 2 6 6 ...
$ n.child   : int 1 6 3 0 0 4 4 3 2 3 ...
$ dep.cons  : int 0 0 0 0 1 0 1 0 1 0 ...
$ scz.cons  : int 0 0 0 0 0 0 0 0 0 0 ...
$ grav.cons : int 1 2 2 1 2 1 5 1 5 5 ...
$ ns        : int 2 2 2 2 2 1 3 2 3 2 ...
$ ha        : int 1 2 3 2 2 2 3 2 3 2 ...
$ rd        : int 1 1 2 2 2 1 2 2 1 2 ...
```

The number of observations (in this case, subjects) and the number of variables are presented in ❶ and ❷. The name of each variable appears in column ❸. Then follows the nature of this variable in ❹. The most frequent types are: integers ("int"), numerical values ("num", in general numbers with decimals) or factors ("Factor", i.e., categorical variables where levels are coded using characters). In ❺, the first 10 values of each variable are displayed; NA stands for missing data (Not Available). For a factor variable, the labels of the different categories appear first, followed by the index of the first 10 observed results.

At this stage, we can note that the first 10 values of variable age are consistent with what is expected for age in adult prisoners (between 20 and 50). The levels of the variable prof (profession) are "farmer," "craftsman," etc., which also appear satisfactory. Variable n.child stands for "number of children." Variables dep.cons and scz.cons (consensus for depression and consensus for schizophrenia, respectively) are coded "0" for "no disorder" and "1" for "disorder." The variable grav.cons stands for "consensus for gravity"; it is a score that ranges from 1 to 7. The variables ns (novelty seeking), ha (harm avoidance), and rd (reward dependence) are ordered categorical variables that range from 1 to 3. So far, all these definitions are consistent with the output obtained from str(mhp.ex). We can now go further and estimate the summary statistics.

The summary() function is the basic instruction to obtain summary statistics with R. For integer or numerical variables, it gives the "five-number" summary (minimum, lower quartile, median, upper quartile, maximum) and the variable mean and number of missing values. For factors (categorical variables), it gives the observed frequencies of the first seven variable levels. If we now consider the following mhp.ex dataset:

```
> summary(mhp.ex)
      age❶                  prof❷              n.child           dep.cons
Min.    :19.00  worker       :228   Min.    :  0.000  Min.    :0.000
1st Qu.:28.00  employee     :136   1st Qu.:  0.000  1st Qu.:0.000
Median :37.00  craftsman    : 91   Median :  1.000  Median :0.000
Mean    :38.94  intermediate: 57   Mean    :  1.572  Mean    :0.392
3rd Qu.:48.00  other        : 31   3rd Qu.:  2.000  3rd Qu.:1.000
Max.    :84.00  (Other)      : 29   Max.    : 14.000  Max.    :1.000
NA's    : 2.00  NA's         :227   NA's    : 26.000
    scz.cons          grav.cons             ns                ha
Min.    :0.00  Min.        :1.000  Min.    :  1.000  Min.    :  1.000
1st Qu.:0.00  1st Qu.     :2.000  1st Qu.:  1.000  1st Qu.:  1.000
Median :0.00  Median      :4.000  Median :  2.000  Median :  2.000
Mean    :0.08  Mean        :3.635  Mean    :  2.058  Mean    :  1.876
3rd Qu.:0.00  3rd Qu.     :5.000  3rd Qu.:  3.000  3rd Qu.:  3.000
Max.    :1.00  Max.        :7.000  Max.    :  3.000  Max.    :  9.000
               NA's        :4.000  NA's    :104.000  NA's    :107.000
          rd
Min.    :  1.000
1st Qu. :  1.000
Median  :  2.000
Mean    :  2.152
3rd Qu. :  3.000
Max.    :  3.000
NA's    :114.000
```

Numerical variables are presented like the variable age in ❶. Factors are presented like the variable prof in ❷. The function summary is a classic but has several drawbacks:

1. It does not estimate the standard deviation.
2. The output takes up too much room on the screen or in a report.

3. The variables are not in rows, with statistics in columns (which would be more convenient if we need to copy results and paste them into a table in a word processor).

The function describe() in the package "prettyR" tackles these different points:

```
> library(prettyR) ❶
> describe(mhp.ex,num.desc = c("mean", "sd", "median", "min",
  "max", "valid.n")) ❷
Description of mhp.ex

Numeric
                mean        sd      median    min    max    valid.n
age❸           38.94      13.26         37     19     84      797❹
n.child        1.572      1.849          1      0     14      773
dep.cons      10.392      0.4884        10     10      7      799
scz.cons      10.080      0.2716        10     10     17      799
grav.cons      3.635      1.648          4     17      7      795
ns             2.058      0.8786         2      1      3      695
ha❺            1.876      0.9125         2      1     9❻      692
rd             2.152      0.8309         2      1      3      685


Factor
           farmer    craftsman    manager    intermediate    employee
prof            5           91         24              57         136
Percent      0.63        11.39          3            7.13       17.02
           worker        other         NA
prof          228           31        227
Percent     28.54         3.88      28.41
mode = worker    Valid n = 572
```

The prettyR package is not accessible by default. It must therefore be installed (the last paragraph of Section 1.4 explains how to install a new package) and then called using the function library() ❶. A given package must be called only once per session.

It is possible to tailor the describe() function according to particular needs. Here ❷, the mean, standard deviation, median, minimum, maximum, and number of available observations have been chosen. Concerning the variable age in ❸, the mean age is about 39 with a standard deviation of 13. If the distribution of age is normal, this means that about two-thirds of the detainees are between the ages of 26 and 52. The minimum age is 19 while the maximum is 84; there are only two missing data (valid.n = 797 in ❹). For the temperament variable "harm avoidance" (ha❺), it can be noted that the maximum ❻ is 9, while this variable should be coded 1 (low), 2 (average), and 3 (high), as for novelty seeking (ns) and reward dependence (rd). This is obviously an error and it must be corrected. The number of missing data for ns, ha, and rd is rather large (about 100, as valid.n = 695, 682, 685, respectively). This is possibly explained by the fact that these variables were assessed at the end of the interview, which could have been terminated prematurely due to the numerous constraints peculiar to prison life.

2.2 Summary Statistics in Subgroups

In a few words: What are the average levels of job satisfaction in male and female employees? What are the percentages of female subjects in subgroups defined by their job? Several R functions can be used to produce cross-tabulations of this type, depending on the nature of the variable analysed and on the layout of the table obtained. The function table() can be used to cross-tabulate two categorical variables, while the function by() is suited for estimating the mean values of numerical variables across levels of a categorical variable.

In Practice: Let us imagine that we are interested in the distribution of the temperament variable "novelty seeking" (named ns and coded "1" for low, "2" for average, and "3" for high) across each level of the variable prof (profession). The function table() gives

```
                  ❶              ❷                      ❸
> table(mhp.ex$prof, mhp.ex$ns, deparse.level = 2,
              ❹
    useNA = "ifany") mhp.ex$ns❺
mhp.ex$prof❺         1         2         3    <NA>❻
farmer               3         1         1       0
craftsman           32        25        18      16
manager             12         1         7       4
intermediate        19        16        13       9
employee            39        28        47      22
worker              76        45        75      32
other                8         6        13       4
<NA>❻               60        35       115      17
```

The two variables profession❶ and novelty seeking❷ appear in an instruction that includes

1. The dataset name (here "mhp.ex")
2. The character "$"
3. The variable name (here, "prof" and "ns")

This can appear somewhat cumbersome; it is in fact an efficient guard against errors in data manipulation. Indeed, several datasets can be used within the same R session, and these datasets can have variables with the same name. The requirement to specify both the dataset name and the variable name eliminates possible ambiguity that could appear if only the variable name was used. The instruction deparse.level = 2❸ provides in ❺ the names of the variables in rows and columns. It can be noted that in the syntax of the function table(), the first variable❶ gives the rows in the table while the second variable❷ gives the columns. The instruction useNA❹ is used to produce rows and columns (❻) corresponding to missing values.

The function table() can also be used in the following way:

```
            ❶                                    ❷
table("Profession" = mhp.ex$prof, "Novelty seeking" = mhp.ex$ns,
   useNA = "ifany")
                  Novelty seeking❷
Profession❶          1       2       3    <NA>
   farmer            3       1       1       0
   craftsman        32      25      18      16
   manager          12       1       7       4
   intermediate     19      16      13       9
   employee         39      28      47      22
   worker           76      45      75      32
   other             8       6      13       4
   nojob            59      33     115      14
   <NA>              1       2       0       3
```

Here, "Profession"❶ and "Novelty seeking"❷ are used to label rows and columns.

The function table() estimates how many observations correspond to each combination of levels, and not the percentages of subjects that correspond to these combinations. When percentages must be estimated, the function prop.table() can be useful:

```
            ❶
> options(digits = 3)
      ❷
> tab <- table(mhp.ex$prof, mhp.ex$ns, deparse.level = 2,
   useNA = "ifany")
                  ❸
> prop.table(tab,1)
                        mhp.ex$ns
mhp.ex$prof         1        2        3     <NA>
   farmer      0.6000   0.2000   0.2000   0.0000
```

```
craftsman        0.3516   0.2747   0.1978   0.1758
manager          0.5000   0.0417   0.2917   0.1667
intermediate     0.3333   0.2807   0.2281   0.1579
employee         0.2868   0.2059   0.3456   0.1618
worker           0.3333   0.1974   0.3289   0.1404
other            0.2581   0.1935   0.4194   0.1290
<NA>             0.2643   0.1542   0.5066   0.0749
```

The option concerning digits❶ is designed to specify the number of digits that will be displayed. The results obtained from the use of the function table() are stored in a new object named tab❷. The option "1"❸ in prop.table() means that percentages are computed by rows (for a given row, the sum of all the percentages is equal to 1). The option "2" can be used for percentages computed by columns:

```
> prop.table(tab,2)
                              mhp.ex$ns
mhp.ex$prof           1         2         3        <NA>
   farmer          0.01205   0.00637   0.00346   0.00000
   craftsman       0.12851   0.15924   0.06228   0.15385
   manager         0.04819   0.00637   0.02422   0.03846
   intermediate    0.07631   0.10191   0.04498   0.08654
   employee        0.15663   0.17834   0.16263   0.21154
   worker          0.30522   0.28662   0.25952   0.30769
   other           0.03213   0.03822   0.04498   0.03846
   <NA>            0.24096   0.22293   0.39792   0.16346
```

By default (if neither option "1" or "2" is used), the percentages are estimated from the whole sample and their sum is equal to 1:

```
> prop.table(tab)
                              mhp.ex$ns
mhp.ex$prof           1         2         3        <NA>
   farmer          0.00375   0.00125   0.00125   0.00000
   craftsman       0.04005   0.03129   0.02253   0.02003
   manager         0.01502   0.00125   0.00876   0.00501
   intermediate    0.02378   0.02003   0.01627   0.01126
   employee        0.04881   0.03504   0.05882   0.02753
   worker          0.09512   0.05632   0.09387   0.04005
   other           0.01001   0.00751   0.01627   0.00501
   <NA>            0.07509   0.04380   0.14393   0.02128
```

If the aim is to compute the means of ns, ha, and rd across the levels of prof, the function by() produces

```
                              ❶                           ❷
> by(subset(mhp.ex, select = c(ns, ha, rd)), mhp.ex$prof,
       ❸              ❹
  mean, na.rm = TRUE)
mhp.ex$prof: farmer
   ns         ha         rd
   1.6        1.8        2.4
---------------------------
mhp.ex$prof: craftsman
   ns         ha         rd
  1.813      1.613      2.162
---------------------------
mhp.ex$prof: manager
   ns         ha         rd
  1.750      2.200      2.474
---------------------------
mhp.ex$prof: intermediate
   ns         ha         rd
  1.875      1.750      2.250
---------------------------
mhp.ex$prof: employee
   ns         ha         rd
  2.070      1.843      2.113
---------------------------
mhp.ex$prof: worker
   ns         ha         rd
 1.995       1.938      2.159
---------------------------
mhp.ex$prof: other
   ns         ha         rd
  2.185      1.667      2.000
```

The fact that means are estimated is specified in ❸. The variable "prof" appears in ❷ while the variables "ns", "ha", and "rd" are selected from the dataset mhp.ex using the function subset() in ❶ (see Section 9.2 for details on the use of subset()). The instruction na.rm should be used to make it clear that variables with missing data should be removed before any computations.

In the package "Hmisc", the function summary() can also be used to present descriptive statistics in subgroups of subjects:

```
> library(Hmisc)
            ❶              ❷                    ❸
> summary(prof ~ ns + ha + rd, method = "reverse",
          ❹
  data = mhp.ex)
```

```
Descriptive Statistics by prof
```

	N	farmer (N=5)	craftsman (N=91)	manager (N=24)	intermediate (N=57)	employee (N=136)
ns❺ 1	695	60% (3)	43% (32)	60% (12)	40% (19)	34% (39)
2		20% (1)	33% (25)	5% (1)	33% (16)	25% (28)
3		20% (1)	24% (18)	35% (7)	27% (13)	41% (47)
ha : 1	691	60% (3)	56% (42)	25% (5)	54% (26)	47% (54)
2		0% (0)	27% (20)	30% (6)	17% (8)	22% (25)
3		40% (2)	17% (13)	45% (9)	29% (14)	31% (36)
rd : 1	685	20% (1)	23% (17)	11% (2)	21% (10)	28% (32)
2		20% (1)	38% (28)	32% (6)	33% (16)	33% (38)
3		60% (3)	39% (29)	58% (11)	46% (22)	39% (45)

The first argument of summary() is a "formula." To the left of "~"❶ appears the name of the categorical variable that determines the subgroups. To the right of "~"❷ are the names of the variables that are to be described. More precisely, for each level of variable prof❶, the frequencies and number of subjects corresponding to each level of each variable in ❷ are displayed. If a variable in ❷ is an integer or is numerical and has more than 10 levels, then quartiles are presented instead of frequencies and numbers of subjects.

2.3 Histograms

In a few words: Mean, median, or standard deviation extract essential pieces of information from data, but cannot represent the actual distribution of a given variable in a precise way. Graphical representations are often unavoidable for this reason, histograms in particular.

One difficult point in constructing a histogram lies in determining the number of bars. If there are too few bars, the representation is not informative enough; if there are too many, the overall shape of the distribution is lost. The function histogram() automatically determines the number of bars and generally yields informative representations.

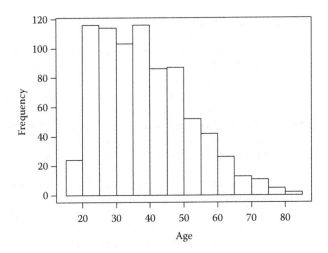

FIGURE 2.1
Histogram of prisoners' age. The y-axis indicates the number of subjects in each bin. The distribution is skewed, with a floor effect around 20 years corresponding to the official age limit of 18 years for being sent to French prisons for adults.

Some authors have pointed to limitations inherent in histograms (Fox 2002); instead, they suggest using a "density curve," which is a smooth, more sophisticated estimate of the distribution of a numerical variable. The R function density() can be used for this purpose.

Examination of a histogram makes it possible to determine whether the distribution is roughly normal (like a bell curve), if it is skewed, if it has several peaks, a ceiling or a floor effect (i.e., a large number of subjects registering the lowest or the highest value, respectively), or if there are outliers (observations that are numerically distant from the remainder of the data).

In Practice: Let us now see how to obtain a graphical representation of the distribution of prisoners' age in the MHP study (Figure 2.1). Calling the function histogram() gives

```
        ❶              ❷                    ❸
> hist(mhp.ex$age, xlab = "Age", ylab = "Frequency",
                ❹
    main = "Histogram of age")
> box()❺
```

The variable name is in ❶, and the labels of the x- and y-axes are in ❷ and ❸, respectively. The instruction main❹ corresponds to the general title of the representation. It may be useful to colour the bins; this can be done, for example, by adding the option ", col = "grey"" after ❹. The instruction box()❺ draws a box around the histogram.

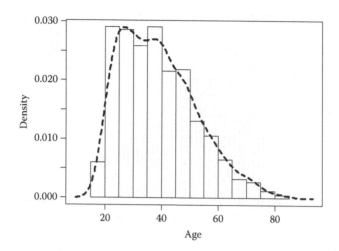

FIGURE 2.2

Histogram of prisoners' age with non-parametric density estimate (dashed curve). The curve is smoother than the outline of the histogram. The y-axis can be interpreted as a *relative* frequency, not an *absolute* frequency.

To add a density estimate for the distribution to the plot (Figure 2.2), the following instructions are used:

```
    ❶
> dest <- density(mhp.ex$age, na.rm = TRUE)
> int <- hist(mhp.ex$age, plot = FALSE)
                                              ❷
> hist(mhp.ex$age, xlim = range(int$breaks, dest$x),
                          ❸
  ylim = range(int$density, dest$y), xlab = "Age",
                ❹                     ❺
  ylab = "density", freq = FALSE, main = "")
    ❻               ❼           ❽
> lines(dest, lty = 2, lwd = 2)
> box()
```

The density obtained from the function density()❶ is stored in the object dest; the histogram is stored in int; and ❷, ❸, and ❹ are used to ensure the compatibility of the scales of the density estimate curve and the histogram. The instruction main = "" is used to remove the general title. In ❻, the density is drawn; ❼ and ❽ specify the type of line (lty = 2 corresponds to a dashed line) and its thickness (a larger number for lwd generates a thicker line). In Figure 2.2 it can be seen that the density curve is smoother than the histogram. A histogram is, however, a classic procedure and is thus often preferred.

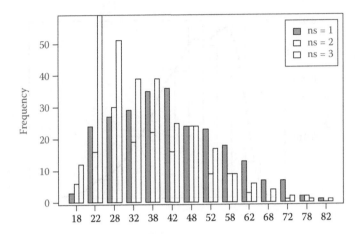

FIGURE 2.3
Multiple histograms of prisoners' age according to three levels of novelty seeking (ns = 1, 2, 3). Subjects with a high level of novelty seeking tend to have a more skewed distribution with younger ages.

It can be useful to represent several histograms on the same diagram, for example the histogram of age for the three levels of novelty-seeking ns = 1, 2, and 3 (Figure 2.3). In the package "plotrix," the function multhist() is dedicated to this task. Here we obtain

```
> library(plotrix)
          ❶                              ❷
> x <- list(mhp.ex$age[mhp.ex$ns == 1],
                      ❸
  mhp.ex$age[mhp.ex$ns == 2], mhp.ex$age[mhp.ex$ns == 3])
> multhist(x, ylab = "n",
  main = "Age according novelty seeking", axis.lty = 1,
                              ❹
  legend.text = paste("ns =", 1:3))
> box()
```

First, a list❶ comprising three elements is required: a variable that gathers the ages of prisoners with a low level of novelty-seeking (mhp.ex$ns == 1❷, see Section 9.2 for details concerning subgroup selections); a variable corresponding to prisoners with a moderate level of ns❸, and a variable corresponding to those with a high level of ns. The legend of the multiple histograms is detailed in ❹. The instruction 1:3 corresponds to the values 1, 2, and 3, and the function paste() can be used to paste strings of characters, so that paste("ns =", 1:3) gives the following result:

```
> paste("ns =", 1:3)
[1] "ns = 1" "ns = 2" "ns = 3"
```

Here, this instruction provides the three labels inside the box: "ns = 1", "ns = 2", and "ns = 3".

2.4 Boxplots

In a few words: Figure 2.3 is interesting because it helps grasp the pattern of the distribution of a given variable in several groups of subjects. It has, however, one drawback: The representation is rather difficult to read, possibly because it contains too much information. A popular alternative to multiple histograms is called the "boxplot" or the "box-and-whisker plot." It is more synthetic than a histogram, and more visual and intuitive than the mere collation of summary statistics such as mean, median, standard deviation, etc.

In Practice: A boxplot graphically represents the median of a variable❶, its first❷ and third❸ quartiles, and its smallest❹ and largest❺ values after having discarded the outliers❻ (an outlier is an observation that is arbitrarily considered as unusually small or large). For instance, if we consider the distribution of prisoners' age according the three levels of novelty seeking, ns = 1, 2 or 3 (Figure 2.4), we have

```
> boxplot(age ~ ns, data = mhp.ex, ylab = "age", xlab = "ns",
  main = "Age according novelty seeking")
```

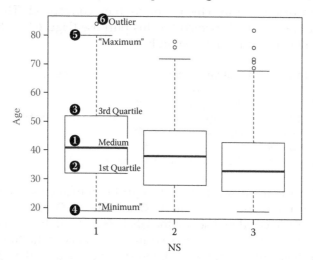

FIGURE 2.4
Boxplot representing the distribution of prisoners' age according to the three levels of novelty seeking (1 for low, 2 for average, 3 for high). The diagram is less informative but more synthetic than the histogram presented in Figure 2.3.

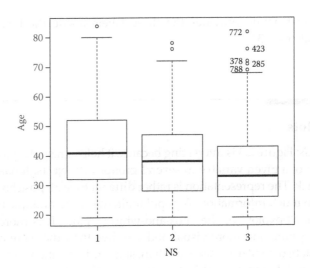

FIGURE 2.5
Identification of outliers using the function `indentify`.

As mentioned above, the notion of outlier is somewhat arbitrary (Balakrishnan and Childs 2001). The minimum❹ and maximum❺ suggested by R are therefore open to discussion; they correspond, respectively, "to the most extreme data points that are no more than 1.5 times the inter-quartile range lower than the first quartile or higher than the third quartile."

Even if these so-called outliers should be considered cautiously, it can be useful to determine to which subjects they correspond. The function `identify()` can be implemented for this purpose (Figure 2.5). It is somewhat difficult to use, but can be helpful:

```
> boxplot(age ~ ns, data = mhp.ex, ylab = "age", xlab = "ns",
  main = "Age according novelty seeking")
> numsubj <- 1:length(mhp.ex$age) ❶
> x <- rep(3, length(mhp.ex$age[mhp.ex$ns == 3])) ❷
> y <- mhp.ex$age[mhp.ex$ns == 3] ❸
       ❹      ❺  ❻                    ❼
> identify(x, y, labels = numsubj[mhp.ex$ns == 3])
```

After the classic use of the function `boxplot()`, the function `identify()`❹ is used with three arguments: the x- ❺ and the y- ❻ coordinates of the points that are detected and so are the labels ❼ that will be used to characterise the outliers. If we are interested in the outliers of the third boxplot, the y-coordinates of the points of interest correspond to the age of subjects with a high level of novelty seeking. This corresponds to the syntax "y <- mhp.ex$age[mhp.ex$ns == 3]" in ❸. Concerning the x-coordinates, the `boxplot()` routine assigns "1" to the x-coordinates of subjects corresponding to the first boxplot (ns = 1), "2" to the subjects corresponding to the second boxplot (ns = 2), and "3" to the last boxplot (ns = 3). This explains the definition of x in ❷ that, using the function `rep()`, generates a vector with as many "3s" as there are subjects in ❸ (`rep()` is a function that "repeats" a number or a vector a certain number of times). Finally, to label the outliers, a vector `numsubj` is created in ❶; it associates to each subject the number of its corresponding row in the data file.

After entering the instruction `identify()` in the R console, each outlier in the third boxplot is pointed to with the mouse and the left mouse button is clicked. The line number of the corresponding prisoner then appears immediately. Once all interesting points have been selected, the "stop" button in the R graphics window is selected. The function `identify()` can be used in the same manner with many other graphical functions.

2.5 Barplots

In a few words: Barplots can be used to represent the distribution of a categorical variable, ordered or not. Barplots are graphically similar to histograms; however,

1. The number of bars is exactly determined from the number of levels for the variable that is represented (it is not arbitrarily determined by the software); and

2. The bars are separated to show explicitly that the variable is not continuous.

In Practice: It is frequent in questionnaires to find a succession of questions concerning the same topic and having a comparable response pattern (e.g., satisfaction concerning a list of products). It can be interesting to obtain a graphical representation of the distribution of the answers to these questions in a single diagram; this can be done easily with R (Figure 2.6). For example, in the MHP study, if we consider the three variables of temperament—novelty seeking (ns = 1, 2, 3), harm avoidance (ha = 1, 2, 3), and reward-dependence (rd = 1, 2, 3)—we can use the following instructions:

```
                ❶
> temp <- c("ns", "ha", "rd")
     ❷              ❸  ❹
> par(mfrow = c(1, 3))
          ❺                           ❻            ❼
> for (i in temp) barplot(table(mhp.ex[, i]), main = i,
                 ❽              ❾
    ylab = "n", space = 0.8, col = "white")
```

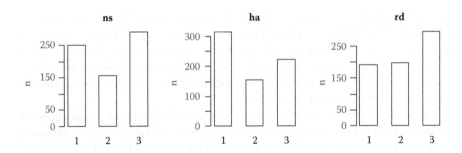

FIGURE 2.6
Barplots representing the distribution of three ordered variables.

In ❶, the vector that contains the name of the variables to be represented is defined. The instruction par❶ is used to split the R graphic windows into three parts: one vertical slot❸ × three horizontal slots❹. To automatically generate the barplot for each temperament variable, a loop is used in ❹. The instruction "i in temp"❺ means that the index i will take all possible values in temp, that is, "ns", "ha,", and "rd". In ❻, there will therefore be three calls to the barplot() function, with three corresponding barplots. The variable name to be presented in each plot is specified in ❼. The instruction space❽ is used to specify the space between the bars; the colour of the bars is defined in ❾, and by default it is grey.

2.6 Pie Charts

In a few words: Pie charts are a popular way of graphically representing the distribution of a categorical variable. They are, however, regularly criticised by the experts (e.g., Spence 2005). The main argument is that the human brain is likely to have difficulty comparing the surface areas of several slices of pie; the comparison of bar length in a barplot could be more intuitive and more reliable. This is likely to be especially true concerning the three-dimensional pie charts that are provided by most spreadsheet programs. These representations should not be used because, depending on the position of the slice in the pie, two slices associated with comparable frequencies can have widely different surface areas. Pie charts can, however, be useful when the objective is to compare the size of each slice with the whole pie.

In Practice: The variable prof (profession) has seven levels: "farmer," "craftsman," "manager," "intermediate," "employee," "worker," and "other." A pie chart for this variable can be obtained (Figure 2.7) from the function pie(table()):

                              ❶   ❷   ❸   ❹
```
pie(table(mhp.ex$prof), col = gray(seq(0.1, 0.85, length = 7)))
```

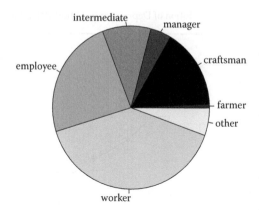

FIGURE 2.7
Pie chart representing the distribution of the variable "profession".

The shading of the slices is determined in ❶. A 0-1 grey scale is used, the lightest shade is in ❷ and the darkest in ❸. There are seven shades❹ because there are seven professions. Pure black (0) and white (1) are discarded to avoid overly strong contrasts.

2.7 Evolution of a Numerical Variable across Time (Temperature Diagram)

🖋 *In a few words*: In a cohort study, the same questions can be asked at different times, and it is often useful to represent the evolution of these variables across time. The progress of each subject is sometimes represented by a broken line. For most studies based on questionnaires, a diagram of this sort is, however, illegible due to the large sample size. Summary statistics are therefore necessary and the mean plus/minus one standard deviation from the mean is often employed.

In Practice: The MHP study is not a cohort study, so that in this instance we are considering a different dataset. The dataset "repdat" contains the Hamilton Depressive Rating Scale (HDRS) scores measured in 186 depressed patients interviewed at the beginning of hospitalisation (t = 0), and after 4, 7, 14, 21, 28, 42, and 56 days. This dataset has the following structure:

```
> str(repdat)
'data.frame':  1053 obs. of 3 variables:
 $ NUMERO : int   96  96  96  96  96  96  96  96 157 157 ...
 $ VISIT  : int    0   4   7  14  21  28  42  56   0   4 ...
 $ HDRS   : int   34  26  12   7   5   1   1   1  27  19 ...
```

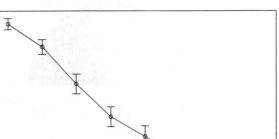

Evolution of Depression in a Short Potential

Evolution of a numerical variable across time (temperature diagram). The sample mean ± 1 standard deviation is represented at each time.

The eight evaluations performed on a given subject are not in separate variables/columns. Instead, there is only one HDRS variable, and each visit (not each patient) corresponds to a particular row. A file of this type is sometimes called a "normalised" file, while the other option (one line per patient with eight HDRS variables) is called a "denormalised" file.

The library "gplots" and its function plotmeans() can very easily produce a temperature diagram (Figure 2.8) when the dataset is normalised. This yields

```
> library(gplots)
              ❶                    ❷
> plotmeans(repdat$HDRS ~ repdat$VISIT, gap = 0, barcol = "black",
  xlab = "Number of days", ylab = "Score of depression",
  main = "Evolution of depression in a cohort of patients")
```

The main instruction is a formula that determines the "y" variable to be represented❶ across time❷. The other options involve the graphical aspect of the representation. It could be argued that Figure 2.9 is potentially misleading. Indeed, a reader might notice that patients' scores are close to the x-axis at day 56 (the end of the study) so that they likely to be no longer symptomatic. This is, in fact, erroneous. The HDRS has its minimum at 0, but the y-axis does not cross the x-axis at y = 0, but around 6. It is possible to deal with this issue by requiring the y-axis to cross the x-axis at y = 0 when calling plotmeans() (Figure 2.9):

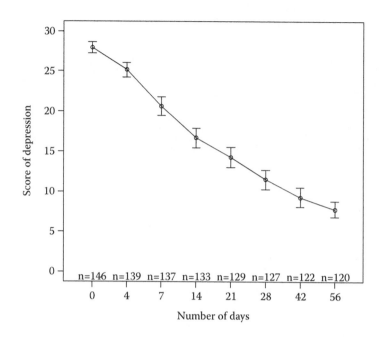

FIGURE 2.9
Evolution of a numerical variable across time. The y-axis crosses the x-axis at y = 0, which is the minimum HDRS depression score. This representation is less misleading than Figure 2.8.

```
> plotmeans(repdat$HDRS ~ repdat$VISIT, gap = 0, barcol = "black",
  xlab = "Number of days", ylab = "Score of depression",
  main = "Evolution of depression in a cohort
                              ❶
  of patients", ylim = c(0, 30))
```

The instruction ylim = c(0, 30)❶ stipulates that the y-axis should range from 0 to 30.

3

Description of Relationships between Variables

If means, medians, standard deviations, and histograms are the basis of a statistical analysis plan, the core of scientific method is the study of relationships between variables. Binary variables are frequent in questionnaire surveys, and odds-ratios and relative risks are parameters that can provide useful quantification of the strength of an association between a binary outcome and a binary predictor. For two numerical variables, the correlation coefficient can be used to represent the strength of the linear component of their dependency. We have seen that one feature of questionnaire studies concerns the large number of variables that are measured and the complexity of their patterns of association. Multidimensional exploratory methods can be used to find the most salient correlations and to comprehensively represent the network of relationships that exist among the variables: hierarchical clustering and principal component analysis (PCA) are essential tools in this context.

3.1 Relative Risks and Odds Ratios

In a few words: Binary variables are frequently an essential part of datasets obtained from questionnaire surveys. This is because it is easy to answer a "yes/no" question, or because it is often comfortable to think in terms of "presence/absence" or "true/false" (although questions as simple as "being single or not", "unemployed or not", or "having bowel cancer or not" are not as binary as they appear at first sight). When a binary variable is an outcome (e.g., being a teacher or not) and when another binary variable is likely to influence this outcome (this can be termed a "risk factor"; for instance, the interviewee's mother or father was a teacher), it is often interesting to quantify the strength of association between the two. The relative risk and the odds-ratio are used for this purpose.

The relative risk (RR) is defined as the probability of reaching the outcome in the presence of the risk factor, divided by the probability of reaching the outcome in the absence of this risk factor. Formally, if the data are as follows,

	Outcome	
	Yes	No
Exposed to risk factor	a	b
Not exposed	c	d

then

$$RR = \frac{\dfrac{a}{(a+b)}}{\dfrac{c}{(c+d)}}$$

As opposed to the odds-ratio (OR) is not easy to interpret

$$OR = \frac{\dfrac{\text{"Outcome = yes" inexposed}}{\text{"Outcome = no" inexposed}}}{\dfrac{\text{"Outcome = yes" in not exposed}}{\text{"Outcome = no" in not exposed}}}$$

$$OR = \frac{\dfrac{a}{b}}{\dfrac{c}{d}}$$

In other words, there are *OR* more times the occurrence of an outcome versus no outcome for the exposed group than for the non-exposed group.

Relative risks and odds-ratios are positive, between 0 and infinity. When they are equal to 1, the outcome and the risk factor are independent. When they are close to zero or infinity, the two variables are highly dependent. If the outcome is a rare occurrence (e.g., below 5% or 10%) or very frequent (over 90% or 95%), then the odds-ratio is close to the relative risk. Indeed, if "*a* is very much below *b*" and "*c* is very much below *d*," then "$b \cong a + b$" and "$c \cong c + d$" so that, from the two formulae above, we have $RR \cong OR$.

While the odds-ratio is less intuitive than the relative risk, it does have two advantages:

1. It is compatible with logistic regression models, powerful statistical tools that will be considered in Section 5.2.

2. It can be estimated in population-based samples such as those in case-control studies. This point is not obvious, but has important implications. An example will help to clarify it.

Consider a study that looks for the association between burnout and gender among school teachers. About a thousand teachers are interviewed in different schools (we call this a "population sample study") and the following data are obtained:

	Burnout	
	Yes	No
Female	75	600
Male	15	300

Thus we have

$$RR = \frac{\dfrac{75}{(75+600)}}{\dfrac{15}{(15+300)}} = 2.33$$

$$OR = \frac{\dfrac{75}{600}}{\dfrac{15}{300}} = 2.5$$

Because the prevalence of burnout is low (around 9%), the odds-ratio (2.5) is close to the relative risk (2.33). Now consider a less costly study design, in which the requirement is to find 90 burnout teachers and 90 controls (this design corresponds to a "case-control study"). Imagine that the gender in the samples of cases and controls is distributed exactly as it was distributed in the population sample study above, that is,

	Burnout	
	Yes	No
Female	75	60
Male	15	30

We now have

$$RR = \frac{\dfrac{75}{(75+60)}}{\dfrac{15}{(15+30)}} = 1.66$$

$$OR = \frac{\dfrac{75}{60}}{\dfrac{15}{30}} = 2.5$$

As anticipated, the odds-ratios are the same in the two situations while the relative risks differ substantially ($RR = 2.33$ versus $RR = 1.66$).

In Practice: Imagine that the objective is now to estimate the strength of association between the outcome "depression in prison" and the risk factor "the prisoner has a high level of harm avoidance" (harm avoidance is a temperament trait). The variable ha is coded 1, 2, and 3 for, respectively, low, average, and high levels of harm avoidance. First we need to define the binary variable ha.b "the prisoner has a high level of harm avoidance: yes(1)/no(0)":

```
                              ❶
> mhp.ex$ha.b <- ifelse(mhp.ex$ha > 2, 1, 0)
> ❷table(mhp.ex$ha.b, mhp.ex$ha, deparse.level = 2,
   useNA = "ifany")
                  mhp.ex$ha
mhp.ex$ha.b      1      2    ❸3    <NA>
     0         315    154     0       0
     1           0      0   222       1
   <NA>          0      0     0     107
```

The function ifelse()❶ is used to create the variable ha.b. If ha > 2 (this corresponds here to ha = 3 for "high"), then ha.b = 1, otherwise ha.b = 0. In ❷, the function table() is used to check that the instruction ❶ correctly coded variable ha.b. In ❸, all subjects for whom ha = 3 verifies ha.b = 1 and all subjects for whom ha = 1 or 2 verifies ha.b = 0.

The function twoby2() in the package "Epi" estimates the odds-ratio and the relative risk between ha.b (high level of harm avoidance) and dep.cons (consensus of junior and senior investigators on depression). A practical point is worth noting here: the labelling of the outcome and the reference exposure. Here, "1" means that a prisoner is depressed (reference for the outcome) and that he has a high level of harm avoidance (reference for the risk factor). Unfortunately, twoby2() considers that the smaller values (here "zero") are the references, so that this function will spontaneously estimate the odds-ratio and the relative risk between "not being depressed" and "not having a high level of harm avoidance." This therefore must be taken into consideration:

```
> library(Epi)
                     ❶                          ❷
> twoby2(1 - mhp.ex$ha.b, 1 - mhp.ex$dep.cons)
2 by 2 table analysis:
------------------------------------------------------------------
Outcome   :  0❸
Comparing :  0❹ vs. 1
            ❻
        0     1    P(0)    95% conf. interval
  0   126    96   0.5676    0.5016     0.6312
❺1  131   338   0.2793    0.2406     0.3217
```

```
                              95% conf. interval
            Relative Risk:  2.0320❼     1.6884    2.4455
        Sample Odds Ratio:  3.3865❽     2.4262    4.7267
Conditional MLE Odds Ratio:  3.3797❾    2.3923    4.7926
    Probability difference:  0.2882     0.2101    0.3626

            Exact P-value:  0
        Asymptotic P-value:  0
```

The first variable❶ appearing in twoby2() is the exposure; the second variable ❷ is the outcome. It can be noted that the two variables are converted so that "0" becomes "1" and vice versa. The reference outcome is specified in ❸; it is "0" and corresponds now to "depressed." The two levels of exposure (risk factor) that are compared are presented in ❹ ("0" corresponds now to a high level of "harm avoidance"). The two-by-two table crossing exposure (in rows❺) and outcome (in columns❻) is then displayed. The relative risk is in ❼ and the odds-ratio in ❽. It can be noted that because depression is not rare here (about 37%), the relative risk is somewhat different from the odds-ratio (2.03 versus 3.38). In ❾ there is a second estimate of the odds-ratio. This can be useful when there are very few outcomes or exposed subjects. Basically, ❽ should be preferred because it is more usual.

3.2 Correlation Coefficients

In a few words: A correlation coefficient* quantifies the strength of the association between two numerical variables (e.g., named X and Y). It is a statistical tool that is essential in psychometrics, a discipline devoted to the construction and validation of composite scores, which are frequent in questionnaire surveys (Chapter 7 deals specifically with psychometric methods).

Correlation coefficients essentially concern monotonic, linear relationships. The relationship between X and Y is monotonic if Y increases (or decreases) when X increases. This can be seen, for example, between satisfaction with life and income, or between a woman's stature and her partner's stature. More precisely, a correlation coefficient expresses how close the x_i and the y_i values are to a straight line when they are represented on a Cartesian plot.

A correlation is a number between –1 and 1. When equal to 1, the x_i and y_i values are on a straight line and we have $Y = a + b \times X$ (with $b > 0$: the relationship is positive). It nearly always corresponds to a situation where the same variable is presented in two different ways (e.g., a height in inches

---

* In this book, by default, "correlation coefficient" stands for "Pearson's correlation coefficient."

and a height in metres). When a correlation is equal to -1, the x_i and y_i values are also on a straight line, but $b < 0$ (so that when X increases, Y decreases). If X and Y are independent, the correlation is equal to 0. The opposite is true only in some circumstances (i.e., when the two variables are jointly normal). Unfortunately, it is impossible to interpret the magnitude of a correlation when it is different from 1, 0, and -1. Some authors (Cohen and Cohen 1975) have proposed that a correlation greater than 0.5 should be considered "large," "medium" between 0.3 and 0.5, and "small" below 0.3. This is consistent with the conscious or unconscious representations of most researchers in the field of human and social sciences. However, such guidelines have definitely no scientific basis (Cohen 1988). Some statisticians argue that the square of the correlation coefficient (sometimes called the coefficient of determination and noted R2 or R^2) can be interpreted as the proportion of variability that the two variables X and Y share, in other words the amount of information they have in common. This interpretation has been criticised on the basis of some interesting arguments (O'Grady 1982; Ozer 1985).

In Practice: In human beings, age is likely correlated with the number of children. Indeed, an older person has had more opportunities to have children. Or older people may belong to a generation for which the expected number of children was larger. What is the strength of this correlation? In the MHP study, it is possible to estimate the correlation coefficient between age (age) and n.child (number of children). The function cor.test() can be used for this:

```
> cor.test(mhp.ex$age, mhp.ex$n.child)

    Pearson's product-moment correlation

data: mhp.ex$age and mhp.ex$n.child
t = 15.9305, df = 771, p-value < 2.2e-16
alternative hypothesis: true correlation is not equal to 0
95 percent confidence interval:
  0.4426556 0.5488913
sample estimates:
  cor
0.4976374❶
```

The result is in ❶ and it is close to 0.5.

3.3 Correlation Matrices

In a few words: During the exploratory phase of the statistical analysis of a questionnaire, it may be interesting to have a table that collates the correlation coefficients of a large number of pairs of variables. A table of this type is called a correlation matrix. It can help identify meaningful relationships that were not considered *a priori*. Correlation matrices raise two questions:

1. Is it possible to consider correlations between all types of variables (numerical, binary, or ordered)?
2. When there is missing data concerning a given variable, is it also necessary to discard the observation for the pairs of variables with no missing data?

Concerning the type of variable that can be used in a correlation matrix, one thing is clear: Non-ordered categorical variables with more than two levels (like "profession" or "country") should be discarded. Apart from that, there is no clear consensus. In short, correlation coefficients can be computed between any numerical, ordered, or binary variables. Caution is, however, required in the interpretation of the results that are obtained. For instance, it is formally possible to compute a correlation coefficient between two binary variables (this is called a phi coefficient). However, these coefficients tend to underestimate the strength of the association between the two variables: Due to numerical constraints, they cannot be close either to 1 or −1 (Cox and Wermuth 1992).

Concerning the question of missing data, there are at least two options. The first is to discard all subjects who have at least one piece of missing data in one of the variables in the correlation matrix. In this instance, if there are many variables and a significant proportion of missing data, the final dataset is likely to comprise very few subjects. The second option is to discard a subject if data are missing for at least one of the two variables considered in each correlation. This option is the so-called "pairwise" correlation matrix. It has the advantage of maximising the amount of information that can be used. However, not all coefficients are estimated from the same sample so that the mathematical properties of the correlation matrix are not optimal. In practice, this second option is often used when the first does not make sense. Of course, there is always the possibility of imputing missing data, but this raises specific problems that are discussed in detail in Section 6.6.

In Practice: In our data file mhp.ex, the only variable that cannot be included in a correlation matrix because of its coding is prof (profession). From the function cor() it is possible to obtain the correlation matrix for the remaining variables:

❶
```
> quanti <- c("age", "n.child", "dep.cons", "scz.cons", "grav.cons",
  "ns", "ha", "rd")
```
❷   **❹**   **❸**
```
> round(cor(mhp.ex[, quanti], use = "pairwise.complete.obs"), digits = 3)
```

	age	n.child	dep.cons	scz.cons	grav.cons	ns	ha	rd
age	1.000❾	0.498❺	-0.093	-0.021	-0.127	-0.223	-0.027	-0.001
n.child	0.498	1.000❾	0.003	-0.003	-0.057	-0.159	0.004	-0.023
dep.cons	-0.093	0.003	1.000❾	0.094	0.454❻	0.109	0.256❽	0.089
scz.cons	-0.021	-0.003	0.094	1.000❾	0.318❼	0.022	0.081	-0.004
grav.cons	-0.127	-0.057	0.454	0.318	1.000❾	0.154	0.230	0.019
ns	-0.223	-0.159	0.109	0.022	0.154	1.000❾	0.081	0.071
ha	-0.027	0.004	0.256❽	0.081	0.230	0.081	1.000❾	0.119
rd	-0.001	-0.023	0.089	-0.004	0.019	0.071	0.119	1.000❾

In ❶, the variables that will be included in the correlation matrix are selected. By default, all correlations are presented with numerous decimals so that the matrix takes up a lot of space and is difficult to read. This is why the function round()❷ is used with digits = 3 ❸, which specifies the number of decimal places retained in the output. As explained above, and because there are many missing values for several variables, the pairwise correlation matrix is computed❹.

The results show that except for the correlations between number of children and age❺, gravity and depression❻, and gravity and schizophrenia❼, all correlations are below 0.3. This should of course be taken cautiously because many of these variables are binary or are ordered with three levels. All correlations appear twice❽, once above and once below the diagonal of 1❾.

It can be frustrating in this output not to have the p-values corresponding to the test that each of these correlations is null. This is possibly a good thing: There is no temptation to engage in a pointless fishing expedition. Furthermore, because only one of the variables has a normal distribution, the validity of these tests is questionable (see Section 4.6 for details about statistical tests concerning correlation coefficients). The function rcor.test() in the "ltm" library nevertheless offers a correlation matrix of this type:

```
> library(ltm)
> rcor.test(mhp.ex[, quanti], use = "pairwise.complete.obs")
```

	age	n.child	dep.cons	scz.cons	grav.cons	ns	ha	rd
age	*****	0.498	-0.093	-0.021	-0.127	-0.223	-0.027	-0.001
n.child	<0.001	*****	0.003	-0.003	-0.057❶	-0.159	0.004	-0.023
dep.cons	0.009	0.936	*****	0.094	0.454	0.109	0.256	0.089
scz.cons	0.554	0.942	0.008	*****	0.318	0.022	0.081	-0.004
grav.cons	<0.001	0.113❷	<0.001	<0.001	*****	0.154	0.230	0.019
ns	<0.001	<0.001	0.004	0.561	<0.001	*****	0.081	0.071
ha	0.478	0.914	<0.001	0.033	<0.001	0.034	*****	0.119
rd	0.986	0.554	0.020	0.920	0.627	0.065	0.002	*****

❸upper diagonal part contains correlation coefficient estimates lower diagonal part contains corresponding p-values

As explained in ❸, the correlation between gravity and number of children is in ❶ (−0.057), while the corresponding p-value is in ❷ (p = 0.113).

3.4 Cartesian Plots

In a few words: Correlation coefficients capture only the linear part of the relationship between two numerical variables. To interpret these coefficients, it is therefore necessary to assess to what extent the relationship is linear. Cartesian plots (or X,Y plots) are essential for this purpose. Because they are often informative, they should be used without restriction, even if the space they take up most often precludes their inclusion in the final version of a report or paper.

In Practice: In Section 3.2 we estimated the correlation coefficient of age with number of children. The instruction plot can be used to obtain a Cartesian plot of these two variables:

```
          ❶                   ❷                   ❸
> plot(mhp.ex$age, mhp.ex$n.child, xlab = "Age",
                       ❹
  ylab = "Number of child")
```

The "x" variable❶ is given first, followed by the "y" variable❷ and optionally the labels of the x-axis❸ and the y-axis❹. This Cartesian plot can be disappointing for at least two reasons:

1. Because all prisoners with the same age and number of children will correspond to the same point, there are somewhat fewer than 797 points while there are 797 prisoners for whom age is available.

2. The graph is not easy to interpret. Obviously the dispersion is considerable and if it does appear that "number of children" increases with "age," this is not completely convincing.

To deal with the first reason above, it is possible to "jitter" the variables, that is, to add some noise to each determination of "age" and "number of children" so that each prisoner will have his own pair of coordinates. To deal with the second reason above, in Figure 3.1 a regression line can be added (the line that gives the linear tendency of the evolution of "number of children" according to "age"). A smooth regression curve can also be represented. This type of curve is called a "spline"; it can be useful in detecting non-linearity in the relationship between the two variables of interest. Section 6.1 deals in more detail with this latter point.

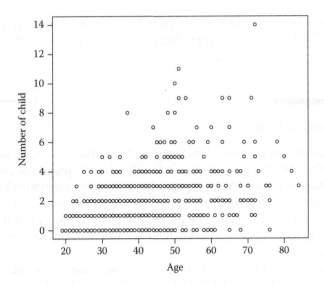

FIGURE 3.1
Cartesian plot of "Age" with "Number of children." Because all prisoners of the same age and having the same number of children will correspond to the same point, there are somewhat fewer points than the 797 prisoners for whom age is available. This can be misleading.

```
         ❶                          ❶
> plot(jitter(mhp.ex$age), jitter(mhp.ex$n.child),
    xlab = "Age (jittered)", ylab = "Number of child (jittered)")
         ❷                                        ❸
> abline(lm(n.child ~ age, data = mhp.ex), lwd = 2, lty = 2)
      ❹
> nona <- !(is.na(mhp.ex$n.child) | is.na(mhp.ex$age))
         ❺
> lines(lowess(mhp.ex$age[nona], mhp.ex$n.child[nona]), lwd = 2)
```

The instruction plot() is used as in the previous example, the only difference being the use of the function jitter()❶ applied to the x- and y-coordinates. The two functions abline() and lm()❷ are used to represent the regression line (drawn with abline()) obtained from a linear model (estimated with lm()). The instructions lwd and lty❸ are used to determine the width or thickness (lwd) of the regression line and its type (dotted line when lty = 2). The variable nona❹ extracts the prisoners who have no missing data for "age" and "number of children." The regression curve (spline) is displayed by way of the instruction lines(lowess())❺.

Figure 3.2 is certainly more informative than Figure 3.1. We can see here that most of the prisoners are aged between 20 and 30 and have no children. One prisoner is about 73 years old and has 14 children. There is a positive linear trend between the two variables and the spline curve does not show any major deviation from linearity (i.e., the curve is rather straight).

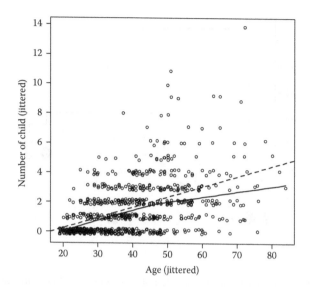

FIGURE 3.2
Cartesian plot of age versus number of children. The coordinates are "jittered" (some noise has been added) so that each prisoner has his own pair of coordinates. The dotted line is a linear regression and the continuous line a spline (i.e., a regression curve).

3.5 Hierarchical Clustering

In a few words: In Section 3.3, a correlation matrix was estimated from the eight numerical, ordered, or binary variables in the "mhp.ex" dataset. From these 27 correlation coefficients, it is not so easy to apprehend the pattern of associations between the eight variables. A possible explanation for this difficulty derives from the fact that, from a geometrical point of view, this correlation matrix can be interpreted as giving distances between eight points in an eight-dimensional space, so that it is likely to be impossible for human beings to have a reliable intuition of a space of this nature.

To tackle this problem, statisticians have developed specific tools that aim to extract the relevant information from the correlation matrix and to present it in a comprehensible manner. One of these tools is known as hierarchical clustering. It represents variables on a kind of tree: The closer these variables are on the branches, the more correlated they are. Let us now see the principle behind this approach.

Let us consider a correlation matrix (Figure 3.3a) for five variables and imagine that these variables can be represented by points on the plane so that the distance between two points represents the correlation between the corresponding pair of variables (the closer the points, the more correlated the variables).*

---

* From a mathematical point of view, the distance $d = 2(1 - r)^{1/2}$, where r is the correlation coefficient.

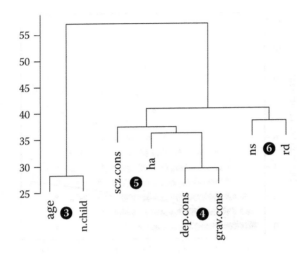

FIGURE 3.3
Hierarchical clustering of a set of variables. Variables joined by a direct pathway in the lower part of the diagram are the most strongly correlated.

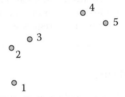

First, the objective is to spot the two points that are closest (they correspond to the two variables that have the largest correlation coefficients)—here points "2" and "3." In a visual representation, we symbolize this by joining two vertical segments named "2" and "3" at a distance d1 on the Y-axis corresponding to the distance d1 of point "2" from point "3" on the plane (Figure 3.3b).

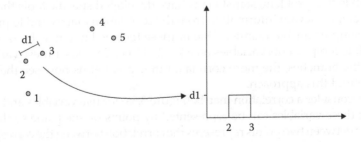

Then we summarize points "2" and "3" by their centre (point "6") and again look for the two closest points: this time points "4" and "5." We complete the representation by joining two new vertical segments at a distance d2 that corresponds to the distance between points "4" and "5" (Figure 3.3c).

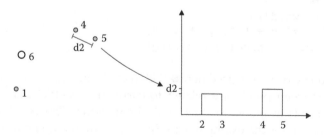

Points "4" and "5" are summarized by their centre: point "7." Then we continue with points "6" and "1" at a distance of d3 (Figure 3.3d).

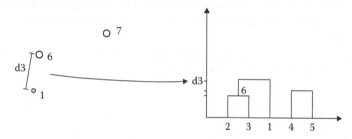

And finally we obtain Figure 3.3e.

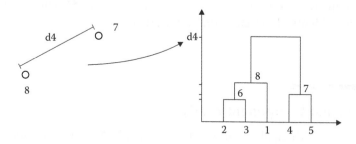

In this diagram, we can now see that variables 2 and 3 are the most correlated, then variables 4 and 5. Variable 1 can then be aggregated to the cluster formed by variables 2 and 3.

In Practice: Because hierarchical clustering uses distances associated to correlation coefficients, this method can be used with numerical, ordered, or binary variables, with the precautions detailed in Section 3.3 concerning correlation matrices. After removing the variable "profession" from the data file mhp.ex ("profession" is the only variable that is neither numerical, ordered, nor binary), the function hclust() is called and we obtain

```
                            ❶
> cha <- hclust(dist(t(scale(mhp.ex[, quanti])))),
            ❷
  method = "ward")
> plot(cha, xlab = " ", ylab = "",
  main = "Hierarchical clustering")
```

The function hclust() is used here❶ with dist(t(scale())) to ensure that distances between points are related to correlation coefficients. There are several methods available for hierarchical clustering. The Ward method❷ has been chosen for the example here. Several methods can be tried successively to ensure that the partition obtained is robust.

In Figure 3.3 we can see that age and number of children❸, and depression and gravity❹, are the two pairs of variables that are most strongly correlated. Schizophrenia and harm avoidance can then be aggregated to the cluster "depression and gravity"❺. Novelty seeking and reward dependence are on their own❻.

The function heatmap() combines the symbolic representation of a correlation matrix with a hierarchical clustering:

```
> obj <- cor(mhp.ex[, quanti], use = "pairwise.complete.obs")
                                                    ❶
> heatmap(obj, col = gray(seq(1, 0, length = 16)))
```

In Figure 3.4 dark squares are associated with strong correlations and white squares with weak correlations (on a 16-degrees scale)❶. The hierarchical clustering is identical to that found in Figure 3.3.

3.6 Principal Component Analysis

In a few words: Principal Component Analysis (PCA) is a very general tool, and we consider here only the question of the graphical representation of a correlation matrix (Hills 1969). In the previous section, we saw that, from a geometrical point of view, variables in a correlation matrix correspond to points in a high-dimensional space. More precisely, these points are on the "unit hypersphere" of a high-dimensional space (Lebart, Morineau, and Piron 1995) (a "hypersphere" is the generalisation to an n-dimensional space of a sphere or a circle in a three- or two-dimensional space). Furthermore, the distances between the points on the hypersphere are directly related to the correlations between the variables. Hence, if we could "see" the points on the unit hypersphere, we would be able to apprehend the patterns of

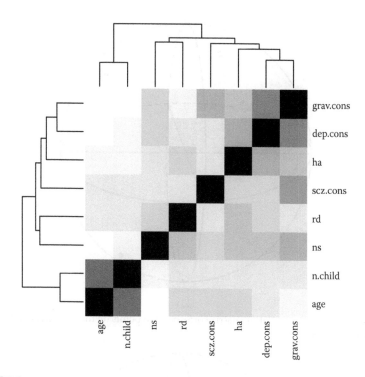

FIGURE 3.4

Symbolic representation of a correlation matrix. Correlation coefficients are characterized by a grey scale. A hierarchical clustering of the variables is also produced.

associations between the corresponding variables. Unfortunately, this is not cognitively achievable.

PCA resolves this dilemma by projecting the points on a plane with minimal distortion. Let us see this in a theoretical example. Consider five points on a three-dimensional sphere as in Figure 3.5.

The objective is to find a projection of these points onto a plane that distorts the original representation as little as possible. A projection of this type is shown in Figure 3.6.

Now, to fully apprehend the patterns of associations between the variables, it is sufficient to look at the plane on which the points are projected (see Figure 3.7).

An interesting property is that if two points on the hypersphere are close and if they are close to the plane, then (1) their projections will also be close and (2) the projections will be close to the circle, which corresponds to the intersection of the hypersphere with the plane. In Figure 3.7 it can be seen that the two points in ❶ are close to one another and close to the circle; it is thus possible to infer that the original points on the hypersphere are also close and this means that the two corresponding variables are positively and rather highly correlated.

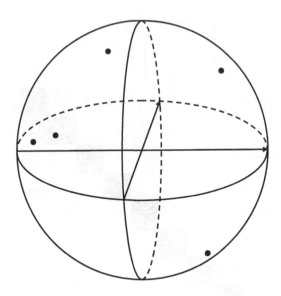

FIGURE 3.5
Variables involved in a correlation matrix correspond to points on the "unit hypersphere" of a high-dimensional space.

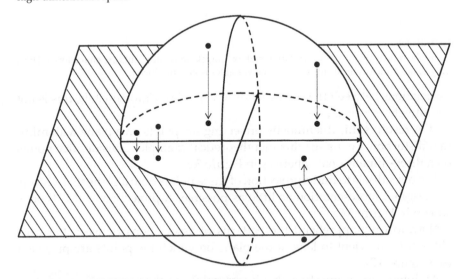

FIGURE 3.6
PCA consists of a "faithful" projection of the points onto a plane.

More generally, the following can be demonstrated (Figure 3.8). When two points are close to one another and close to the circle, then the two underlying variables are positively correlated ❶. When two points are close to the circle and form a right angle with the origin (O), then the two underlying variables are uncorrelated ❷. When two points are diametrically opposed

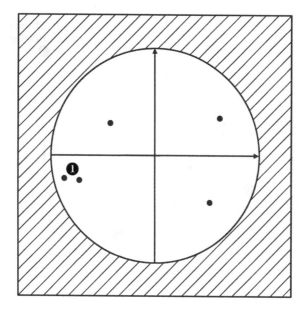

FIGURE 3.7
When points are close to the circle, their relative position can be interpreted in terms of their relative position on the sphere. Here, because the two points in ❶ are close to the circle, it is possible to say that the two corresponding variables are positively correlated.

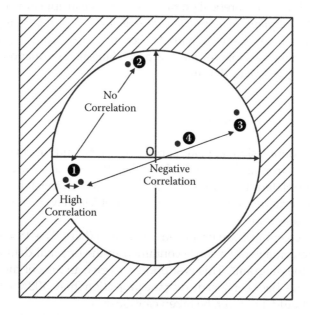

FIGURE 3.8
Rules for the interpretation of a PCA diagram.

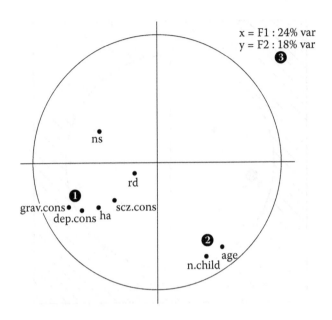

x = F1 : 24% var
y = F2 : 18% var

FIGURE 3.9
PCA plot graphically representing the correlation matrix of eight variables.

and close to the circle, then the two underlying variables are negatively correlated ❸. When a point is close to the origin O, nothing can be said ❹.

In Practice: From a theoretical point of view, PCA can appear very academic and difficult to implement. In practice, the reverse is true; it is straightforward to construct a PCA diagram with R. For instance, from the numerical, ordered, or binary variables of the "mhp.ex" dataset, the function mdspca() in the library "psy" gives (Figure 3.9):

```
> library(psy)
> mdspca(mhp.ex[, quanti])
```

Gravity and depression appear correlated in ❶ and uncorrelated with number of children and age in ❷. These last two variables are likely to be positively correlated (they are close and close to the circle). It is unwise to interpret the relative position of the other variables; they are too close to the centre of the diagram.

In ❸, two percentages are proposed. Their sum (here, 24 + 18 = 42%) represents the "percentage of variance contained in the original data that is represented by the diagram". This is supposed to give an idea of the amount of information that has been lost during the projection (here, 100% − 42% = 58%). This result should, however, be taken cautiously.

PCA is an appealing method that is easy to use and simple to interpret. In practice, however, it is often disappointing, especially when too many variables are considered. Indeed, it is tempting to select a large number of variables (why not all of them?), and the PCA diagram is then expected to generate a "comprehensive picture" of all these data. Unfortunately, when more and more variables are added, the points overall move closer to the centre of the diagram until it is no longer interpretable. As a general rule, it can be considered that a PCA diagram is actually informative when fewer than 10 variables are used, and rarely more than 15.

Because PCA geometrically represents a correlation matrix, this method can, theoretically, be applied to numerical, ordered, or binary variables. However, as has been said previously, results obtained from binary or ordered variables should be viewed cautiously. There are methods that are close to PCA but specific to categorical variables (such as Multiple Correspondence Analysis (Greenacre and Blasius 2006)), although they are rather more delicate to use.

3.7 A Spherical Representation of a Correlation Matrix

In a few words: If variables involved in a correlation matrix can be represented by points on a hypersphere, why does PCA project them onto a plane and not onto a three-dimensional sphere? The representation could be more informative and just as easy to interpret. This is provided by a "spherical representation of a correlation matrix"(Falissard 1996). The plot obtained can be interpreted like a PCA. When two points are close, the two underlying variables are positively correlated; when two points form a right angle with the origin, the two underlying variables are uncorrelated; and when two points are diametrically opposed, the two underlying variables are negatively correlated.

In Practice: The function sphpca() in the library "psy" is also easy to use (Figure 3.10):

❶   ❷

```
> sphpca(mhp.ex[, quanti], cx = 0.9, v = 55)
```

The instruction cx = ❶ is used to modify the length of the names given to the variables. The instruction v = ❷ rotates the sphere so that all the points can appear on the same side. A rotation of 55° in the vertical plane (v = 55❷) has been chosen in Figure 3.10 after several trials.

As in Figure 3.9, age and number of children are positively correlated (the two points are close in ❸), independent of depression and gravity (❸ and ❹ approximately form a right angle with the centre of the sphere), and negatively correlated with novelty seeking (❸ and ❺ are diametrically opposed). The spherical representation is more informative than a PCA

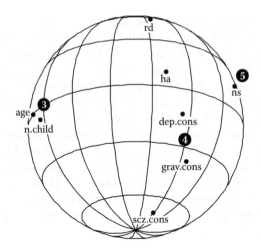

FIGURE 3.10
Spherical representation of the correlation matrix of eight variables. The interpretation of the relative positions of the variable points is similar to a traditional PCA plot.

plot. It has, however, one drawback: There is no way to determine, for a given variable, if it is well represented or not in relation to its correspond-ing point (this was possible with PCA, using the distance between the point and the unit circle).

3.8 Focused Principal Component Analysis

In a few words: In many questionnaires, certain variables have a central role and are considered outcome or dependant variables, while certain other variables are expected to explain these outcomes and they are predictors or explanatory variables. This dichotomy is not compatible with the diagrams provided by hierarchical clustering or PCA, where all the variables have the same status. "Focused" PCA tackles this problem. One variable is considered an outcome and the others are predictors. The representation is "focused" on the outcome, so that the distances between the predictors and the outcome are a faithful representation of their correlations. The relative positions of the predictors can give an idea of their correlations and can be interpreted as in a PCA diagram.

In Practice: The function fpca() in the library "psy" can perform focused PCA. In the mhp.ex data file, if the variable grav.cons (consensus assess-ment of the gravity of a prisoner's psychiatric state) is the outcome and the other variables the predictors, then we obtain (Figure 3.11):

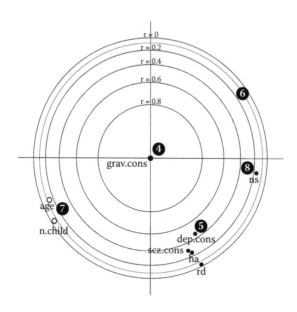

FIGURE 3.11

Focused PCA for the eight numerical variables in the file mhp.ex. The focus is on the variable "gravity." Correlations between this variable and the predictors are faithfully represented. The correlations between the predictors are interpreted as in a PCA.

```
                 ❶
> response <- "grav.cons"
                 ❷
> explanatory <- c("age", "n.child", "dep.cons", "scz.cons",
  "ns", "ha", "rd")
> fpca(data = mhp.ex, y = response, x = explanatory,
                 ❸
  partial = "No")
```

The outcome is determined in ❶, and the predictors in ❷. In the fpca() call, the instruction partial = "No" ❸ is required to obtain a representation close to a PCA for the predictors.

In Figure 3.11, the focus is on the variable "gravity" (in the centre of the diagram❹). The variable "depression"❺ is the closest to ❶; it is therefore the most strongly correlated with "gravity." The correlation is between 0.4 and 0.6 (the point is inside the two corresponding circles), for example, about 0.43. This correlation is significant at the 5% level because it is inside the dotted circle ❻. The variables "age" and "number of children" are likely positively correlated because the corresponding points are close to each other in ❼. They are likely independent of "depression" because ❼ and ❺ form a right angle with the centre of the circle. They are likely negatively correlated to novelty seeking because ❽ and ❺ are diametrically opposed.

4

Confidence Intervals and Statistical Tests of Hypothesis

When the time comes to claim new findings, and when these findings are likely to influence important decisions, it is essential to determine to what extent they are reliable. Among the possible sources of misinterpretation, mistaking noise for signal is frequent, and potential sources of noise are indeed numerous in questionnaire surveys. One source is the inevitable level of uncertainty in the response that is given to a question. Another arises from the fact that, when conducting a survey, a limited sample of subjects is interviewed, while results tend to be generalized to an underlying population of interest.

The present chapter deals with these issues. They have generated a large corpus of theoretical statistical work, among which confidence intervals and statistical tests of hypothesis are the two main focuses. And although these tools are in daily use, they are still open to epistemological controversy.

4.1 Confidence Interval of a Proportion

In a few words: Imagine that a thousand men have been interviewed and that 16% of them reported not having watched TV the previous Sunday. When the time comes to interpret this result, and if this interpretation has practical implications, it is in general implicitly assumed that the value of 16% can be more or less generalised to a given country, region, or city. Indeed, most often, a survey is not intended to discover something about a sample, but rather to know something about a large population from which the sample is derived. But to what extent is a generalisation of this type actually possible? Confidence intervals are an attempt to answer this question.

Confidence intervals have a clear meaning in at least one situation: when the interviewees are a *random* sample from an underlying population (theoretically, a random sample can be obtained only if a list of all members of the population is available and if all the subjects who are randomly selected from this list agree to participate in the study; this rarely happens in practice). In this situation, a probabilistic model can be developed and used to estimate an interval that contains, with a probability of, for example, 95%, the proportion of interest (the "true" proportion of interest: in our example, the proportion of adult men in a given country who would report not having watched TV the previous Sunday)*.

In some circumstances, the sample is obtained randomly but by way of a more complex process. In the MHP study, prisons were randomly selected from French prisons for men, and then prisoners were randomly sampled in these prisons. In multi-stage sampling of this sort, it is still possible to estimate confidence intervals, but the statistical tools required are more complex than in the previous situation.

Sometimes, there is only a pseudo-randomisation, for example when interviewees are selected "at random" in the street. In this situation, confidence intervals have no rigorous meaning. They are nevertheless still potentially useful because they can be interpreted as rough fluctuation intervals (Schwartz 1986): If the same experiment was repeated on another day in the sample place, or the same day in a different place within the same area, the percentage of men reporting not having watched TV the previous Sunday will be slightly different. If the number of replications tends toward infinity, the 95% confidence interval will contain about 95% of all the observed results. Briefly, in this context, confidence intervals give an idea of the influence of sampling fluctuations on the value of the parameter of interest.

In Practice: One of the main questions behind the MHP study concerns the proportion of subjects with a severe form of schizophrenia in French prisons for men. The variable scz.grav is appropriate to address this question because it is equal to 1 if a prisoner simultaneously has a consensus diagnosis of schizophrenia and a consensus severity score of at least 5 ("markedly ill") and to 0 otherwise. This variable is available in the data frame "mhp.ex2" as are the two variables centre and type.centre2, which correspond, respectively, to the prison to which the prisoner belongs and to the corresponding type of prison ("1" for « maison centrale » (high security units for very long sentences), "2" for « centre de détention » (for longer sentences and/or prisoners with good prospects for rehabilitation), and "3" for « maison d'arrêt » (for remand prisoners and/or short sentences)).

Let us imagine first that the sampling of prisoners was uniformly random with no missing data (in other words, that interviewees were selected at random from the whole population of French male prisoners). The function binom.confint() in the library "confint" can then be used for the estimation of a 95% confidence interval:

---

* Strictly speaking, this definition is not correct because a given interval either contains or does not contain the true prevalence. A better but less intuitive definition of a 95% confidence interval can be set out as follows: If large numbers of samples of this kind are collected and if a confidence interval is computed each time, about 95% of these intervals will contain the true prevalence. For simplicity, we will retain the "not formally exact" definition in the remainder of the book.

❶
```
> y <- na.omit(mhp.ex2$scz.grav)
> library(binom)
                                                                    ❷
> binom.confint(x = sum(y), n = length(y), method = "asymptotic")
                                 ❸        ❹       ❺
       method      x     n     mean    lower   upper
1 asymptotic      53   795   0.0667   0.0493   0.084
```

The function na.omit() is used to discard missing data in the variable scz.grav. There are several methods that can be used to estimate confidence intervals. The option proposed here (method = "asymptotic")❷ is the most classic. The proportion of prisoners with a severe form of schizophrenia is thus 6.7%❸, with a 95% confidence interval [0.049❹,0.084❺].

The classic option chosen in ❷ has rather good statistical properties (D'Agostino, Chase, and Belanger 1988) but is not appropriate for "small" or "large" percentages (if n denotes the sample size and p the percentage, a "small" percentage is traditionally defined by p ≤ 5/n and a "large" one by p ≥ 1 − (5/n); however, this rule should be used cautiously (Brown, Cai, and DasGupta 2001)). There is statistical debate concerning the approach that should be used for "small" and "large" percentages. The option "method = "wilson"" has certain advantages (Agresti and Coull 1989). In this case we obtain

```
> binom.confint(x = sum(y), n = length(y), method = "wilson")
                                 ❻        ❼       ❽
      method      x     n     mean    lower   upper
1   wilson       53   795   0.0667   0.0513   0.0862
```

As expected, ❻ is similar to ❸; the new confidence interval [0.051❼,0.086❽] is comparable to the previous one, which was [0.049,0.084].

As mentioned above, these results are obtained on the hypothesis that prisoners were straightforwardly sampled from the whole population of French prisoners. This is not true in the MHP study. First, prisoners were sampled using a two-stage stratified design: Prisons were randomised first and then prisoners in these prisons were selected. Second, the strata have unequal weights: type 1 prisons (high-security units) are intentionally over-represented, while type 3 prisons (for remand prisoners and/or short sentences) are underrepresented. This implies that the variance used in the usual formula for the confidence interval is not correct and that the MHP study sample is not a definitive picture of French prisons for men. The proportion of prisoners with severe schizophrenia observed in the sample and the 95% confidence intervals computed above are thus biased.

To deal with this issue, the function svyciprop() in the package survey is used:

```
> mhp.ex2$pw[mhp.ex2$type.centre2 == 1] <- 20❶
> mhp.ex2$pw[mhp.ex2$type.centre2 == 2] <- 45
> mhp.ex2$pw[mhp.ex2$type.centre2 == 3] <- 67❷
> mhp.ex2$strat.size[mhp.ex2$type.centre2 == 1] <- 13❸
> mhp.ex2$strat.size[mhp.ex2$type.centre2 == 2] <- 55
> mhp.ex2$strat.size[mhp.ex2$type.centre2 == 3] <- 144
> library(survey)
                         ❹        ❺                ❻
> mhp.survey <- svydesign(id = ~centre, strata = ~type.centre2,
    weights = ~pw, fpc = ~strat.size, data = mhp.ex2)
                         ❼
> res2 <- svyciprop(~scz.grav, mhp.survey, na.rm = TRUE,
              ❽
    method = "beta")
> res2
              ❾       2.5%      97.5%
scz.grav    0.0630    0.0347     0.1
```

First, the weight of each observation is defined (variable mhp.ex2$pw). For instance, a prisoner in the study living in a high-security unit (type.centre2 == 1) represents 20 prisoners❶ because the sampling rate was 1/20 in this instance. The sampling rate was smaller in type 3 prisons, so that each prisoner in the study has a greater weight (here, 67❷). Then, for each observation, the size of the corresponding stratum is defined (variable mhp.ex2$strat.size). For instance, there are 13 high-security units in France❸.

The function svydesign()❹ gathers all the information related to the survey design (unit of randomisation, weights, strata sizes, dataset) in a single object (named "mhp.survey"): The unit that has been randomised is defined in ❺ (here the prison, which corresponds to the variable centre); the variable identifying the stratum is in ❻, followed by the variables weights (pw) and stratum size (strat.size).

Finally, the function svyciprop() is called to estimate the proportion of prisoners presenting severe schizophrenia ❼. The option method = "beta"❽ is used to give conservative confidence intervals even for small percentages. The option corresponding to the classic algorithm is method = "mean." The prevalence is finally presented in ❾; it is 0.063. This is the estimate of the true prevalence of severe schizophrenia in French prisons for men (different from the proportion of subjects having this disorder in the sample, which was 0.0667 in the previous analysis). The 95% confidence interval is given immediately afterwards: [0.0347,0.1].

4.2 Confidence Interval of a Mean

In a few words: From a theoretical point of view, the confidence interval of a mean raises the problems we have already encountered in the previous paragraph concerning confidence intervals of proportions. However, one technical aspect is different. When the sample size is "large enough," the mean of a quantitative variable follows a normal distribution. This property can be used to estimate a confidence interval. Unfortunately, "large enough" is vague and cannot be specified in a practical way. Sometimes a sample size as large as n = 500 may be necessary (Snedecor and Cochran 1989), depending on the shape of the variable studied: The more skewed and further from normality it is, the larger the sample size required. Most questionnaire studies include a sufficient number of interviewees so that this problem is, in general, marginal. If in doubt, a bootstrap procedure can be used instead of the classic procedure; see Section 6.7 for details.

In Practice: Let us now consider that we are interested in the 95% confidence interval of prisoners' age. If the prisoners in the MHP study had been sampled from the whole population of French prisoners, with no missing data, this confidence interval could be estimated from the formula

$$\text{Mean(age)} \pm k \times \text{Standard deviation(age)}/n^{1/2},$$

where $k \cong 1.96$. This formula can be programmed in R with the following syntax:

```
            ❶
> y <- na.omit(mhp.ex2$age)
> m <- mean(y)
            ❷
> n  <- length(y)
> se <- sd(y)/sqrt(n)
                  ❸
> lowci <- m - qt(0.975, n - 1) * se
> upci  <- m - qt(0.025, n - 1) * se
  ❹
> cat("mean: ", m, " 95% CI: ", lowci, upci, "\n")
mean: 38.9 95% CI: 38 39.9
```

The function na.omit()❶ is used to discard all missing data from the variable "age" while the function length()❷ is used to estimate the number of prisoners for whom age is available. The constant k presented above is obtained from the function qt()❸, which computes quantiles of the t distribution. To obtain a printout that is easy to read, the function cat() is used in ❹; it concatenates words (such as "mean:") with results (such as the content of variable "m").

The formula above is not compatible with the hierarchical nature of the sampling process used in the MHP study. A more complex approach must be used to obtain a confidence interval compatible with the design of the study. It uses the "mhp.survey" object constructed in the previous section and the function svymean() in the "survey" package:

```
> res1 <- svymean(~age, mhp.survey, na.rm = TRUE)
> res1
        ❺
        mean       SE
age     37.8     0.95
> confint(res1)
        ❻
        2.5%     97.5%
age     36.0     39.7
```

The mean for age found here in ❺ (37.8) is different from the previous result obtained in ❹ (38.9). This is due to the fact that the sampling in the MHP study over-represents high-security prisons, where prisoners have long sentences and are therefore older on average. The 95% confidence interval is [36.0,39.7]❻; it is wider than the naive interval obtained in ❹ and this was expected: Two-stage random sampling generally leads to estimates that are less reliable than those derived from a basic one-stage random sampling procedure.

4.3 Confidence Interval of a Relative Risk or an Odds Ratio

In a few words: When the strength of association between a binary outcome and a binary predictor is estimated with a relative risk or an odds-ratio, it is helpful to assess to what extent this parameter depends on sampling fluctuations. A simple way to do this is to compute a confidence interval. As explained previously, this confidence interval will require cautious interpretation, depending on the type of randomisation that was used in the constitution of the sample.

In Practice: In Section 3.1 we found that there is an association between a high level of harm avoidance and depression. The relative risk (RR) that quantified the magnitude of this association was $RR = 2.03$ and the odds-ratio $OR = 3.39$. The function twoby2() in the package "Epi" that was used for these estimations also provides "naive" 95% confidence intervals for RR and OR ("naive" because these estimates assume that the sampling is a single-stage random sampling):

```
> library(Epi)
> twoby2(1 - mhp.ex$ha.b, 1 - mhp.ex$dep.cons)
2 by 2 table analysis:
------------------------------------------------------------------
Outcome    : 0
Comparing  : 0 vs. 1
        0     1    P(0)     95% conf. interval
0     126    96   0.5676    0.5016         0.6312
1     131   338   0.2793    0.2406         0.3217

                              95% conf. interval
               Relative Risk:   2.0320   1.6884    2.4455❶
           Sample Odds Ratio:   3.3865   2.4262    4.7267❷
 Conditional MLE Odds Ratio:    3.3797   2.3923    4.7926
       Probability difference:  0.2882   0.2101    0.3626

              Exact P-value:   0
         Asymptotic P-value:   0
------------------------------------------------------------------
```

How can ❶ and ❷ be interpreted? The sample analysed here was obtained from a two-stage randomisation process (randomisation of prisons, then randomisation of prisoners from these prisons). The function twoby2() does not take this into account. Hence it is not possible to interpret ❷ [2.42,4.72] as an interval where there is a 95% probability of finding the "true" odds-ratio (i.e., the *OR* that would be estimated from the whole population of French male prisoners). The interval ❶ is, however, not totally meaningless. First, if one considers that the same percentage of prisoners has been sampled in each prison, then ❶ is an approximation of the 95% confidence interval that corresponds to the population of interest, that is, "population of prisoners who live in the prisons selected to participate in the MHP study" rather than "the whole population of French male prisoners." Second, one can imagine a virtual underlying population of interest that is homothetic to the sample actually selected (homothetic applies, e.g., to the population obtained by duplicating the sample a large number of times). An approach of this sort could appear as pure fantasy. It nevertheless has potential value because it enables us to grasp the potential influence of sampling fluctuations. The question of generalisation of results to a population of interest in which interventions could be planned is, however, no longer relevant here.

If we are indeed interested in the "whole population of French male prisoners," then the appropriate estimate of the odds-ratio and its corresponding 95% confidence interval can be achieved using the function svyglm() from the package "survey":

❶ ❷
```
> res <- svyglm(dep.cons ~ ha.b, design = mhp.survey,
    na.action = na.omit, family = "binomial"❸)
Warning message:
In eval(expr, envir, enclos) : non-integer #successes in a
    binomial glm!
    ❹
> exp(res$coefficients["ha.b"])
ha.b
3.55
        ❺
> exp(confint(res, "ha.b"))
        2.5%    97.5%
ha.b    2.13    5.92
```

The estimations performed by svyglm()❶ use a bivariate logistic regression (family = "binomial"❸), where the outcome is "depression" and the predictor is "high level of harm avoidance"❷. Logistic regressions are statistical models; they are presented in more detail in Section 5.2. The exponential of the regression coefficient❹ is equal to the odds-ratio. Its 95% confidence interval can be obtained from the function confint()❺. The new *OR* (3.55) is close to the naive *OR* (3.39). Its 95% confidence interval [2.13,5.92] is broader than the naive one [2.42,4.72].

4.4 Statistical Tests of Hypothesis: Comparison of Two Percentages

In a few words: Hypothesis tests are statistical tools designed to help investigators control the risk that is taken when making inferences or basing decisions on probabilistic data. While these tools are easy to implement in practice, they are nevertheless conceptually more complex than is generally assumed (Goodman 1993; Lehman 1993). Statistical tests of hypothesis can be presented according two different theoretical frameworks.

The Neyman and Pearson approach is the more classic and academic approach. Here, two hypotheses are considered, and the objective is to find which is more compatible with the data. Consider, for instance, the situation where two proportions p1 and p2 are to be compared; there is, on the one hand, the so-called null hypothesis H0: p1 = p2 (H0 corresponds classically to a *status quo*) and, on the other hand, the alternative hypothesis HA: p1 ≠ p2 (HA is, in general, the hypothesis that the investigator wants to demonstrate). In this situation there are two ways of making a wrong decision: (1) to conclude to HA when H0 is true (the probability of this error is called alpha) and (2) to conclude to H0 when HA is true (the probability of this error is called beta; see Section 4.9 for details concerning beta). Neyman and Pearson have demonstrated a theorem that leads to a series of decision algorithms for the choice of H0 or HA that minimizes beta when alpha is fixed (alpha is usually set at 5%). These decision algorithms have well-known names such as the "chi-square test," the "Student t-test," etc.

The other approach concerning statistical tests uses the well-known "p-values" that pervade most scientific papers and books. Imagine that two proportions p1 and p2 are compared and that the experiment has led to the observations $p1_{obs} = 13\%$ and $p2_{obs} = 21\%$. The idea is to determine to what extent these two observed values are compatible with the hypothesis that p1 = p2. This can be done by estimating the probability p (the "p-value"), a difference being observed "by chance" between $p1_{obs}$ and $p2_{obs}$ that is at least as large as the difference 21% − 13% = 8% found in the experiment. By construction, if p is close to 0, it is unlikely that p1 = p2. There is a general consensus to consider that if p < 5%, the difference between p1 and p2 can be considered "statistically significant" such that p1 ≠ p2.

It can be noted that the statistical procedures developed in the Neyman and Pearson approach (Student's t-test, chi-square test, etc.) also consist, practically speaking, of the computation of a p-value. If p < 5%, then HA is accepted; while if p ≥ 5%, then H0 is accepted. The two approaches presented in the two previous paragraphs can thus appear strictly equivalent. This is, however, not completely true. In the Neyman and Pearson perspective, the p-value has no meaning in itself; it is just a calculation that can lead to a choice between H0 and HA with the risks alpha and beta. This is not the case with the other approach. Although the decision rules are the same in both cases, the p-value indeed has a meaning in the second instance, where the level of evidence is not the same if p = 0.001 or if p = 0.048. This is a statistical debate (Lehman 1993) and if it has no real practical implications, it is not free from consequences as to what level of evidence is obtained when a statistical test is performed.

In Practice: Imagine that one wants to show that the prevalence of severe forms of schizophrenia (variable "scz.grav") is higher in prisoners with high levels of harm avoidance than in the other prisoners. Before performing the chi-square test required to test the equality of these two percentages, the distribution of prisoners between the two variables—"severe form of schizophrenia" and "high level of harm avoidance"—should be described using the function table():

```
> table(mhp.ex2$ha.b, mhp.ex2$scz.grav, deparse.level = 2)
              mhp.ex2$scz.grav
mhp.ex2$ha.b     0        1
           0    448      18❶
           1    204      17❷
```

As expected, the diagnosis of severe schizophrenia is not frequent in the MHP study. There are only n = 35 patients (❶ + ❷). The function chisq.test()

can then be used to perform a chi-square test that compares the prevalence of schizophrenia in the two samples with and without a high level of harm avoidance:

❶

```
> chisq.test(mhp.ex2$ha.b, mhp.ex2$scz.grav, correct = FALSE)

  Pearson's Chi-squared test

data: mhp.ex2$ha.b and mhp.ex2$scz.grav
X-squared = 4.55, df = 1, p-value = 0.03297❷
```

The instruction `correct = FALSE`❶ is required to obtain the non-corrected chi-square test (i.e., the classic test). The corrected chi-square (`correct = FALSE`) is traditionally proposed for small samples or small percentages. In these situations, a Fisher test should be used instead (see below). The "p-value" is in ❷ with p = 0.033. Because p < 0.05, it is legitimate to state that there is a "statistically significant" difference in the prevalence of severe schizophrenia between the prisoners with a high level of harm avoidance and the others.

A chi-square test is valid only if the percentages that are compared are not too close to 0% and 100% (especially if the sample is not large). Certain formal rules have been proposed to operationalise the notion of being "too close" to 0% and 100%. These rules are not easy to use but fortunately they are implemented by the function `chisqtest()` so that a warning appears automatically when the test appears to be invalid:

```
Warning message:
In chisq.test(y, x) : Chi-squared approximation may be incorrect
```

When this message appears, an alternative to the chi-square test should be used, for instance the Fisher exact test that is implemented in R with the function `fisher.test()`. In our example, the following results are obtained:

❶
```
> fisher.test(mhp.ex2$ha.b, mhp.ex2$scz.grav)

  Fisher's Exact Test for Count Data

data: mhp.ex2$ha.b and mhp.ex2$scz.grav
p-value = 0.04066  ❷
alternative hypothesis: true odds ratio is not equal to 1
95 percent confidence interval:
  0.981 4.358
sample estimates:
odds ratio
  2.07
```

The syntax ❶ is similar to that used with the function chisq.test(). The p-value is in ❷; here again, it is below 0.05.

It can be noted that the chi-square test and the Fisher exact test do not take into account the two-level randomisation of the MHP study. This is noteworthy because the definition of a p-value ("the probability of observing 'by chance' a difference that is at least as great as the observed difference") implies theoretically that (1) the phrase "by chance" does indeed correspond to a random process and (2) the algorithm used to compute the probability adequately models this random process. Obviously, the latter condition is not fulfilled here because a chi-square test or a Fisher test considers implicitly that there is a single-level random process. As seen in Section 4.1 and under certain assumptions, it is possible to interpret the above p-values if the underlying population of interest is restricted to the prisoners detained in the prisons that have been selected to participate in the study, or even if the population of interest is a virtual one. If the whole population of French prisoners is preferred, then another routine should be used, such as the function svychisq() in the "survey" package.

```
          ❶              ❷                  ❸
> svychisq(~scz.grav + ha.b, design = mhp.survey)

    Pearson's X^2: Rao & Scott adjustment

data: svychisq(~scz.grav + ha.b, mhp.survey)
F = 3.3989, ndf = 1, ddf = 17, p-value = 0.08274❹
```

The two variables "scz.grav"❶ and "ha.b"❷ are included on the right-hand side of a formula, with no variable to the left of "~". The object "mhp.survey"❸ was presented in Section 4.1; it gathers data with sampling weights and structure. The p-value is 0.083❹; it is larger than 0.05, while the chi-square test and the Fisher test yielded p-values smaller than 0.05.

4.5 Statistical Tests of Hypothesis: Comparison of Two Means

In a few words: The comparison of two means follows the same logic as the comparison of two proportions. The objective here is to assess to what extent the difference observed between two means ($m1_{obs} - m2_{obs}$) is compatible with the equality $m1 = m2$. A p-value can be computed for this purpose; its meaning is similar to that presented in the previous section, that is, the probability of obtaining, by chance, a difference that is at least as large as that observed in the sample.

The procedure used to obtain this p-value is called the Student t-test. The t-test is valid if the two means compared are obtained from a variable that has a normal distribution. If such is not the case, the t-test can still be used if the sample is large enough (a minimum of 30 observations in each group is generally required). In both cases, the variance of the variable of interest should be the same in the two groups that are compared.

In Practice: Is there a difference in age between prisoners who have high levels of the temperament trait "reward dependence" and the others? To answer this question, a t-test can be used; the corresponding R function is t.test():

```
       ❶
> by(mhp.ex2$age, mhp.ex2$ha.b, sd, na.rm = TRUE)
mhp.ex2$ha.b: 0
[1] 13.3❷
------------------------------------------------------------------
mhp.ex2$ha.b: 1
[1] 13.4❸
       ❹                                              ❺
> t.test(mhp.ex2$age~mhp.ex2$ha.b, var.equal = TRUE)

   Two Sample t-test

data: mhp.ex2$age by mhp.ex2$ha.b
t = 1.67, df = 689, p-value = 0.09442❻
alternative hypothesis: true difference in means is not equal
   to 0
95 percent confidence interval:
   -0.314 3.957
sample estimates:
mean in group 0 mean in group 1
   39.5❼ 37.7❽
```

First, the condition of validity of the t-test is verified. The function by() ❶ is used to compute the standard deviation (sd) of age of prisoners who have❸ or do not have❷ high levels of harm avoidance. The results are similar; as a rule of thumb, it is sometimes suggested that no standard deviation should be greater than 1.5 times the other. In the call to the function t.test()❹, it is specified that the two variances are assumed equal in the two groups ❺. Without this option, R proposes a non-standard version of the t-test that gives conservative results when the variances are not equal. The p-value is in ❻; it is above 0.05 so that there is no statistically significant difference in age between the groups. It is useful to note that the mean ages of prisoners with and without a high level of harm avoidance is presented in ❽ and ❼, respectively.

The sample size (the number of prisoners included in the study) is sufficiently large to guarantee that there are more than 30 observations in each
group (see Section 3.1): Normality is thus not required for age. Nevertheless,
it is a good habit to have a look at the distribution of a numerical variable
involved in a t-test. In the present example, this distribution is presented
graphically in Figure 2.2. It appears slightly skewed. If the sample size was
small, the validity of the t-test would thus be debatable and an alternative to
the t-test should be sought. The most classic is the Mann-Whitney test, which
compares the means of the ranks of the ages of prisoners instead of the means
of the ages themselves. The corresponding R function is `wilcox.test()`:

```
                              ❶
> wilcox.test(mhp.ex2$age ~ mhp.ex2$ha.b)

 Wilcoxon rank sum test with continuity correction

data: mhp.ex2$age by mhp.ex2$ha.b
W = 56649.5, p-value = 0.06096❷
alternative hypothesis: true location shift is not equal to 0
```

The syntax of `wilcox.test()` ❶ is similar to the syntax used with `t.test()`.
The p-value is in ❷.

As in the section devoted to the comparison of two percentages, the tests
presented here do not take into account the two-level randomisation of the
MHP study. The corresponding p-values should therefore be used cautiously. If the objective is indeed to make an inference concerning "prisoners' age" and "having or not having a high level of harm avoidance" in the
whole population of French prisoners, then the function `svyttest()` in the
package "survey" should be used:

```
                       ❶
> svyttest(age ~ ha.b, design = mhp.survey)

  Design-based t-test

data: age ~ ha.b
t = -1.1705, df = 16, p-value = 0.2590❷
alternative hypothesis: true difference in mean is not equal
    to 0
sample estimates:
difference in mean
   -1.687168
```

The syntax ❶ is similar to the syntax used in the `t.test()` function. The
p-value is in ❷.

4.6 Statistical Tests of Hypothesis: Correlation Coefficient

In a few words: If somebody throws a die 20 or 30 times with the left hand and another die with the right hand, the correlation coefficient estimated from the two series of results is, in general, close to 0, but not exactly equal to 0. Following the principle introduced in Section 4.4, it is possible to compute a p-value that assesses the likelihood of the estimated correlation being observed "by chance." If the p-value is greater than 0.05, it is said that the correlation is not statistically different from 0.

The test showing that a correlation coefficient is equal to 0 is valid if at least one of the two variables involved in the correlation has a normal distribution. If this is not the case, a non-parametric correlation can be estimated and tested: the Spearman correlation coefficient.

In Practice: In prisoners, does the temperament trait "novelty seeking" decrease with age? To answer this question, a correlation coefficient can be estimated and tested against 0. The function cor.test() can be used for this purpose:

```
> cor.test(mhp.ex2$age, mhp.ex2$ns)

❶Pearson's product - moment correlation

data: mhp.ex2$age and mhp.ex2$ns
t = -6.009, df = 693, p-value = 3.016e-09❸
alternative hypothesis: true correlation is not equal to 0
95 percent confidence interval:
  -0.2920740 -0.1506631
sample estimates:
  cor
-0.2225387❷
```

In ❶, it is specified that a Pearson's correlation be computed. The correlation is in ❷ (−0.22); it is negative so that older age is associated with a lower level of novelty seeking. The p-value is in ❸; it is below 0.05, which shows that the correlation is statistically different from 0.

The variable "novelty seeking" has only three possible values, so it cannot be considered normal. If one considers that the variable "age" has a distribution that is too skewed to be considered normal (Figure 2.2), then the condition of validity of the test is not satisfied. There are two ways of dealing with this situation. The first is to use a bootstrap procedure (see Section 6.7). The second is to estimate and test a non-parametric coefficient such as the Spearman correlation coefficient. The function cor.test() can again be used:

```
> cor.test(mhp.ex2$age, mhp.ex2$ns, method = "spearman"❶)

  ❷Spearman's rank correlation rho

data: mhp.ex2$age and mhp.ex2$ns
S = 68476247, p-value = 2.412e-09❸
alternative hypothesis: true rho is not equal to 0
sample estimates:
  rho
-0.2238768❹
```

```
Warning message:
In cor.test.default(mhp.ex2$age, mhp.ex2$ns,
  method = "spearman") :
  Cannot compute exact p-values with ties >
```

The option method = "spearman"❶ is now added. It is explicit in the output in ❷. The correlation coefficient is in ❹ with the p-value in ❸.

Here again, the p-values obtained with cor.test() do not take into account the two-level randomization of the MHP study. If there is a need to generalize the results to the whole population of French prisoners, then the function svyglm() in the package "survey" should be used:

```
                  ❶              ❷
> summary(svyglm(age ~ ns, design = mhp.survey))

Call:
svyglm(age ~ ns, design = mhp.survey)

Survey design:
svydesign(id = ~centre, strata = ~type.centre2, weights = ~pw,
  fpc = ~strat.size, data = mhp.ex2)

Coefficients:
              Estimate  Std. Error  t value  Pr(>|t|)
(Intercept)   44.9828   2.1373      21.047   4.35e-13    ***
ns            -3.4190   0.7589      -4.505   0.000360❸   ***
---
Signif. codes: 0 '***' 0.001 '**' 0.01 '*' 0.05 '.' 0.1 ' ' 1

(Dispersion parameter for gaussian family taken to be 167.3585)

Number of Fisher Scoring iterations: 2
```

The two variables "age"❶ and "novelty seeking"❷ are grouped in a formula with a "~". The p-value is in ❸. There is no correlation coefficient available.

4.7 Statistical Tests of Hypothesis: More than Two Groups

In a few words: Most comparisons between groups concern only two groups. It happens, however, that more than two groups are to be compared. In this case, each group can be contrasted with all the others and the interpretation of results can appear problematic. For instance, if the mean of a score is compared across three groups A, B, and C, it is possible to obtain the following: "the result in A *is not* statistically different from the result in B," "the result in B *is not* statistically different from the result in C," while "the result in A *is* statistically different from the result in C." There is a way to avoid such disconcerting circumstances: to test the equality of all means in A, B, and C in a global manner. This can be done using a "one-way analysis of variance" (one-way ANOVA).

When percentages will be compared in three or more groups, the chi-square test and the Fisher test can be used, just as in the two-group situation.

Sometimes, percentages are compared in several groups but these groups can be ordered one in relation to the other. This is the case, for example, if the percentage of people with a regular practice of sport is compared across groups A (drink less than one glass of alcohol a week), B (drink more than one glass of alcohol a week, but less than one glass a day), and C (drink more than one glass of alcohol a day). In a situation of this type, the traditional chi-square test can be used, but it will not take into account the fact that group B is somewhere "between" groups A and C. If one hypothesis is that there is a linear trend in sports practice across groups A, B, and C, then there is a more powerful alternative to the chi-square test: the chi-square test for linear trends.

In Practice: Is there an association between the binary variable "severe schizophrenia (1 for yes versus 0 for no)" and the three-level variable "harm avoidance (1 for low, 2 for average, 3 for high)"?

Before engaging in a statistical procedure, let us begin with the two-way table that crosses the two variables:

```
> table(mhp.ex2$ha, mhp.ex2$scz.grav, deparse.level = 2,
    useNA = "ifany")
               mhp.ex2$scz.grav
mhp.ex2$ha      0     1   <NA>
         1    299    14      2
         2    149     4      1
         3    204    17      1
      <NA>     90    18      0
```

The chi-square test of equality for prevalence of severe schizophrenia across the three groups defined by "harm avoidance" = 1, 2, or 3 can be obtained from the chisq.test() function:

```
> chisq.test(mhp.ex2$ha, mhp.ex2$scz.grav, correct = FALSE)

    Pearson's Chi-squared test

data: mhp.ex2$ha and mhp.ex2$scz.grav
X-squared = 5.28, df = 2❶, p-value = 0.07132❷
```

The degree of freedom of the test ❶ is equal to the number of groups minus 1 (this is not only a mere statistical detail; it can be used to verify the number of groups that are compared). The p-value is p = 0.071❷. The validity of this test is automatically checked by R. When a warning appears, the Fisher exact test should be used instead of the chi-square test. The appropriate function is fisher.test() as in the comparison of two percentages.

If it is preferable to take into account the ordered nature of the three levels of harm avoidance, the chi-square test for linear trend should be preferred. The function prop.trend.test() can be used for this purpose:

```
            ❶
> events <- table(mhp.ex2$ha, mhp.ex2$scz.grav)[, 2]
> events
  1    2    3
 14    4   17
            ❷
> trials <- events + table(mhp.ex2$ha, mhp.ex2$scz.grav)[, 1]
> trials
   1    2    3
 313  153  221
> prop.trend.test(events,trials)

    Chi-squared Test for Trend in Proportions

data: events out of trials, using scores: 1 2 3❸
X-squared = 2.3427, df = 1❹, p-value = 0.12595❺
```

The "events" (i.e., numerators of the percentages to be compared) are computed first❶, and then the "trials" (denominators of the percentages) in ❷. The coding of the variable "harm avoidance" is presented in ❸. This is not just a technical detail, as this coding has a numerical impact on the results: All equally spaced codes such as 1, 3, 5 or 0, 1, 2 would provide the same result; conversely, a code such as 1, 2, 4 would lead to something different. It can be noted that the number of degrees of freedom is now equal to 1❹. The p-value is above the significance level because in ❺, p = 0.126.*

---

* In a large sample, the chi-square test for linear trend gives a p-value similar to the p-value provided by the correlation test between the binary variable that defines the percentages and the ordered variable that defines the groups. In this example, the instruction cor.test(mhp.ex2$scz.grav, mhp.ex2$ha) gives p = 0.1262.

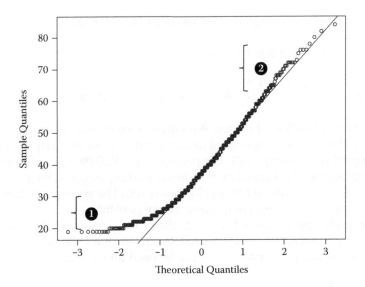

FIGURE 4.1
Normal probability plot (Q-Q plot) for prisoners' ages. If the variable has a normal distribution, all the points should be on the straight line, from bottom left to top right.

Let us now imagine that we are interested in testing a possible difference in age across the three groups of prisoners defined by low, moderate, or high levels of harm avoidance. If these groups are considered independent, the appropriate statistical procedure is a one-way ANOVA. Its conditions of validity are that (1) the variable studied has a normal distribution and (2) its variance is constant across groups. Figure 2.2 gives the histogram and the density curve for prisoners' age; the distribution appears slightly skewed. Some authors argue that it is not easy to assess the normality of a distribution from a simple histogram, and that a specialized tool called a "normality plot" should be preferred (Rutherford 2001). The functions qqnorm() and qqline() (see Figure 4.1) can be used to obtain a representation of this type:

```
> qqnorm(mhp.ex2$age) ; qqline(mhp.ex2$age)
```

If the normality assumption is verified, all the points should be on the straight line. For a given age (y-axis), if a point is above the line, then there are fewer prisoners of this age or a younger age than would be expected if the distribution were normal. This is obviously the case in ❶, where a plausible explanation is that a prisoner cannot be under 18. The same is observed in ❷: The number of older prisoners is larger than expected under a hypothesis of normality (this is perhaps due to long sentences).

This representation is considered one of the most sensitive approaches to assessing normality. It is, however, not easy to understand. As an alternative,

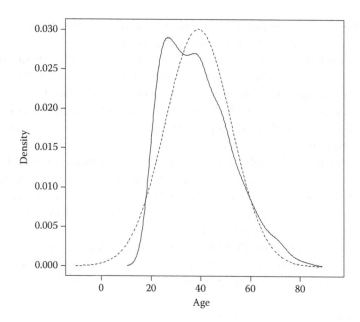

FIGURE 4.2
Density of prisoners' ages superimposed on the distribution of a normal variable with the same mean and standard deviation.

a histogram (or a density curve) superimposed on a normal distribution with a comparable mean and standard deviation is a representation that can be useful (Figure 4.2). The following instructions can be used:

```
> m <- mean(mhp.ex2$age, na.rm = TRUE)
> stddev <- sd(mhp.ex2$age, na.rm = TRUE)
> rangenorm <- c(m - 3.72 * stddev, m + 3.72 * stddev)❶
> density1 <- density(mhp.ex2$age, na.rm = TRUE)❷
> density2 <- dnorm(pretty(rangenorm, 200)❸, mean = m,
  sd = stddev)
> plot(density1, xlim = range(c(density1$x, rangenorm))❹,
  ylim = range(c(density1$y, density2))❺, xlab = "Age",
  main = "")
> lines(pretty(rangenorm, 200)❻, density2❼, lty = 2)
> box()
```

In ❶, the vector rangenorm determines the range of representation for the normal distribution. The density() function is called a first time for variable age❷. The density of the normal distribution is obtained from the function dnorm() on 200 points❸. The x- and y-axes in the diagram should be sufficiently extended to accommodate both density curves; this is the reason why the options xlim❹ and ylim❺ are used. The function lines()

draws the curve corresponding to the normal density function; the x- and y-coordinates of this curve are given in ❻ and ❼.

It is somewhat easier here to see that no prisoner is under 18 and that, as compared to a normal distribution with the same mean and standard deviation, there are possibly more prisoners between 20 and 30 and over 60 years old. To sum up, we can conclude that the distribution of age is not strictly normal; it is slightly skewed. In practice, we will nevertheless consider that the distribution is "sufficiently" normal to perform a one-way ANOVA because this method is robust with respect to non-compliance with the normality assumption, especially when the sample size is large (as is the case in most questionnaire studies) (Rutherford 2001). Some statisticians will not be comfortable with this approximation, and they will therefore need to use a robust approach such as the bootstrap (see Section 6.7).

It is now time to verify that the variance is constant across groups. It is, in general, easier to interpret a standard deviation than a variance (because a standard deviation uses the same unit as the variable itself):

```
> by(mhp.ex2$age, mhp.ex2$ha, sd, na.rm = TRUE)
mhp.ex2$ha: 1
[1] 13.4❶
---------------------------------------------------------mhp.ex2$ha: 2
[1] 13.0❷
---------------------------------------------------------mhp.ex2$ha: 3
[1] 13.4❸
```

The three values are in ❶, ❷, and ❸. They are close, so that it is possible to consider that the variance is constant across groups. As a rule of thumb, the ratio of the largest variance to the smallest should be less than 1.5.

Because the conditions of validity of the one-way ANOVA are now verified, we perform the computation using the lm() function:

```
      ❶              ❷
> res <- lm(age ~ as.factor(ha), data = mhp.ex2)
      ❸
> drop1(res, test = "F")
Single term deletions

Model:
age ~ as.factor(ha)
               Df  Sum of Sq    RSS     AIC   F value   Pr(F)
<none>                         122212   3582
as.factor(ha)   2      1062    123274   3584    2.99    0.0510❹.
---
Signif. codes:  0 '***' 0.001 '**' 0.01 '*' 0.05 '.' 0.1 ' ' 1
```

An ANOVA is, in fact, an elementary linear model; this is why the function lm() is called in ❶. The instruction as.factor()❷ is crucial, and its oversight is a frequent source of error. Indeed, the variable mhp.ex2$ha is coded as a numerical variable (1, 2, and 3) while here it is supposed to index three groups. It therefore needs to be converted into a categorical variable and this is what is provided by the function as.factor()❸. The instruction drop1() is then used to obtain the p-value ❹, which is just above the level of significance: There seems to be no statistically significant association between age and the level of harm avoidance considered as a categorical variable.

4.8 Sample Size Requirements: Survey Perspective

In a few words: If a survey is planned to estimate the proportion of households owning at least two cars, a crucial question is what should be the sample size; in other words, how many households is it necessary to contact to obtain "valid" results? If the households are contacted randomly and if it is hypothesized that there will be no refusals to participate, the sample size is generally obtained from the expected precision of the percentage of interest.

Curiously, before the beginning of the study, the investigators need to guess an estimate of the percentage they are looking for. Here, for example, it will be postulated that about 25% of households own at least two cars. It is also necessary to determine the level of precision required for the percentage. Here, for example, this will be ±5% (more formally, the 95% confidence interval of the percentage of households owning two cars must be as small as [20%,30%]).

When the sampling process involves several steps (as in the MHP study where prisons were selected at random and then prisoners within these prisons), this must be taken into account in the computations.

In Practice: When the MHP study was planned, it was suggested that a prevalence of mental disorder of about 10% should be estimated with a precision of ±2% (so that the 95% confidence interval of the prevalence would be equal to [8%,12%]). Under the hypothesis that the sampling process is random without any clustering or stratification, the sample size can be estimated using the function n.for.survey() in the package "epicalc":

```
> library(epicalc)
                   ❶                ❷
> n.for.survey(p = 0.1, delta = 0.02)

Sample size for survey.
Assumptions:
   Proportion      = 0.1
   Confidence limit = 95%
```

```
Delta                   = 0.02 from the estimate.

Sample size             = 864❸
```

The anticipated prevalence is in ❶ and the required precision in ❷. The resulting sample size is given in ❸.

As noted above, in the MHP study there is a two-level sampling design. This particularity means that the required sample size computed above is not accurate; it is likely under-estimated. To obtain a more acceptable value, it is necessary to guess *a priori* a "design effect." The design effect mathematically translates the fact that two prisoners living in the same prison are likely to be more similar in terms of mental health than two prisoners from two different prisons. A design effect equal to 1 corresponds to an absence of any particular similarity; and the higher the value, the more similar are detainees from the same prison. It is indeed difficult to make a guess about a design effect. Some papers propose a review of the literature concerning particular topics (Rowe et al. 2002; Ukoumunne et al. 1999); however, it is difficult to have really relevant data in most practical situations. Concerning the MHP study, a design effect of 4 was anticipated so that for a 95% confidence interval of [0.06,0.14], the required sample size is

```
                      ❶              ❷              ❸
> n.for.survey(p = 0.1, delta = 0.04, deff = 4)

Sample size for survey.
Assumptions:
    Proportion        = 0.1
    Confidence limit  = 95%
    Delta             = 0.04 from the estimate.
    Design effect     = 4

Sample size           = 864❹
```

The expected prevalence is in ❶, the precision in ❷, and the design effect in ❸. The sample size suggested is 864❹. It was finally reduced to n = 800.

4.9 Sample Size Requirements: Inferential Perspective

In a few words: The main objective of a survey is not necessarily to estimate percentages; it can sometimes aim to estimate the association between two variables (e.g., an outcome and a risk factor). In this situation, the sample size is estimated from the statistical power of the test that will assess the association between the outcome and the risk factor. Formally, following Section 4.4, the statistical power is equal to 1 – beta, where beta is the probability of accepting H0 when HA is true. More simply, the statistical power is equal to the probability that the study will conclude to a significant result, given a certain level of association between the two variables of interest.

When the outcome and the risk factor are binary variables, in addition to the desired power, the investigators need to guess:

1. The level of occurrence of the outcome in the two groups (exposed and not exposed to the risk factor), and
2. The global level of exposure to the risk factor in the total sample.

In Practice: The MHP study was mainly a prevalence study, but it was also designed to find risk factors for psychiatric disorders in the particular context of prison. Therefore the sample size was determined as follows. Let us consider a risk factor that affects 25% of prisoners. Let us also consider that a given psychiatric disorder has a prevalence of 5% in prisoners without the risk factor and of about 12.5% in prisoners presenting the risk factor (the relative risk that associates the disorder to the risk factor is thus equal to 2.5). If we consider that a power of 90% is required to show that an association of this type is statistically significant at the 5% level, then the sample size will be calculated as follows:

```
                ❹              ❸              ❷              ❶
> n.for.2p(p1 = 0.125, p2 = 0.05, power = 0.9, ratio = 3)

Estimation of sample size for testing Ho: p1 == p2
Assumptions:

    alpha =  0.05
    power =  0.9
       p1 =  0.125
       p2 =  0.05
    n2/n1 =  3

Estimated required sample size:

       n1 =  205
       n2 =  615
  n1 + n2 =  820❺
```

The function n.for.2p() in the package "epicalc" is called. It is hypothesized that the number of prisoners without the risk factor is three times greater than the number prisoners with the risk factor❶ (25% versus 75%). The desired power is 0.9❷. The prevalence of the disorder is 0.05 in the population of prisoners without the risk factor❸ and 0.125 in the population of prisoners with the risk factor❹. The corresponding sample size is in ❺; it is 820, which is compatible with the 800 patients finally included. It can be noted here that no design effect is taken into account (which corresponds to a single-step sampling process).

5

Introduction to Linear, Logistic, Poisson, and Other Regression Models

The traditional view of scientific activity is generally the ideal picture of the physicist working in his (her) laboratory. All the forces and constraints potentially interacting with the system can be listed. Thanks to a clever design, they are all maintained in a fixed state—except for two. One is then set at different values and the investigator looks carefully, each time, for the evolution of the second so that a "law" can be postulated that best summarizes all the observations. This is ideal indeed, and very difficult to implement in the area of human and social sciences where questionnaires are most often used. Nevertheless, the fact remains: it is often of great interest to assess the specific effect of "all the forces and constraints potentially interacting with the system." This is basically the objective of regression models. Of course, it is likely impossible, with human beings, to list all "the forces and constraints" or to maintain them in a fixed state. However, it is possible to measure and integrate many of them into a mathematical model so that their respective effects on a given outcome can be assessed. In practice, these mathematical models are never "true" or "exact" but, when designed cautiously and judiciously, they are often very useful.

5.1 Linear Regression Models for Quantitative Outcomes

In a few words: Linear regression is basically dedicated to situations where the outcome Y (also called the response, the dependent variable) is a quantitative variable with a normal distribution.* The predictors X_1, X_2, ..., X_p (also called regressors, explanatory, or independent variables) can be numerical or categorical. Ordered predictors are considered either as numbers or as pure categories. Categorical predictors with more than two levels are converted into a series of binary variables (see Section 6.2 for details). We thus consider from now on that X_1, X_2, ..., X_p are numerical or binary (coded 0 and 1, for instance).

---

* Strictly speaking, it is not necessary for the distribution of Y to be normal; rather, it is the distribution of the residual ε that should be normal. In practice, one situation is frequently associated with the other.

The mathematical model underpinning a linear regression is

$$Y = a_0 + a_1 \times X_1 + a_2 \times X_2 + \ldots + a_p \times X_p + \varepsilon$$

where ε is called a residual; it is assumed to be "pure" random noise with a zero mean.

From this formula it can be noted that if $X_2, \ldots, X_p$ are kept constant and if Y is compared in subjects with $X_1 = C$ to Y in subjects with $X_1 = C + 1$, then

$$Y_{[X1 = C +1]} = a_0 + a_1 \times (C + 1) + a_2 \times X_2 + \ldots + a_p \times X_p + \varepsilon \qquad (5.1)$$

$$Y_{[X1 = C]} = a_0 + a_1 \times C + a_2 \times X_2 + \ldots + a_p \times X_p + \varepsilon' \qquad (5.2)$$

And from Equations (5.1) and (5.2) we obtain

$$Y_{[X1 = C]} - Y_{[X1 = C +1]} = a_1 + (\varepsilon - \varepsilon')$$

Finally, because the mean of $\varepsilon - \varepsilon''$ is equal to 0, a_1 is equal, on average, to the difference in the outcome Y corresponding to a variation of one unit of X_1 when $X_2, \ldots, X_p$ are kept constant. This is exactly what we were looking for: the specific effect on the outcome of a given predictor. It is even possible to test the hypothesis $a_1 = 0$. The corresponding p-value will be interpreted as "the probability of observing 'by chance' a coefficient a_1 at least as large as the one already observed"; in other words, "to what extent the specific association of X_1 with Y can be explained solely by chance."

Like the t-test or the chi-square test, the test of $a_1 = 0$ is valid only under certain conditions: ε must have a normal distribution and a constant variance, ε should be independent of Y, and the X_i; finally, ε should have no pattern of autocorrelation. It is actually difficult, in practice, to be totally confident with all these points. Furthermore, most "occasional" statisticians believe that it is complex and time-consuming to deal with them. In consequence, it is not rare that the question is totally forgotten. This is not coherent. If we are strict with the validity of a t-test or a chi-square test, then the same thoroughness should be applied to regression models. This is all the more relevant because R provides very efficient functions to assess the condition of the validity of these models in a few minutes: If the model is stored in the object "mod," it is sufficient to use the three instructions plot(mod, 2), plot(mod, 1), and plot(mod, 4).

Some authors support the use of a coefficient of determination R2 in multiple linear regressions. Here, R2 is defined as the square of the correlation coefficient between the Y outcomes and the predicted values $a_0 + a_1 \times X_1 + a_2 \times X_2 + \ldots + a_p \times X_p$. We will not use the R2 here for two reasons: (1) its relevance has been criticised (O'Grady 1982; Ozer 1985) and (2) it is useful essentially for the purpose of predicting new outcomes on the basis of the available data, and this is not so frequent in questionnaire surveys.

In Practice: Let us imagine that, in the MHP study, the aim is to determine the variables that are most strongly associated with the duration of the interview. More precisely, it is postulated *a priori* that certain variables should be associated with a longer duration: age (for cognitive reasons), the presence of a psychiatric disorder (because there will be more questions in the clinical part of the interview), a personality disorder (subjects could have difficulty engaging in a relationship with the interviewers), the occurrence of trauma during childhood (subjects could be inclined to talk about it), and the type of prison (a high-security prison could be reluctant about long interviews). A linear regression of the outcome "duration of interview" explained by all the above variables is thus planned.

The first part of this analysis is a return to the very basic description of the data. Using the function describe() of the package "prettyR" we obtain

```
> describe(mhp.mod[, c("dur.interv", "scz.cons", "dep.cons",
  "subst.cons", "grav.cons", "char", "trauma", "age",
  "f.type.centre")], num.desc = c("mean", "sd", "median",
  "min", "max", "valid.n"))
```

Numeric❶

	mean	sd	median	min	max	valid.n❷
dur.interv	61.89❸	19.67	60	0❹	120	749
scz.cons	0.0801	0.2716	0	0	1	799
dep.cons	0.3917	0.4884	0	0	1	799
subst.cons	0.2653	0.4418	0	0	1	799
grav.cons	3.635	1.648	4	1	7	795
char	1.514	0.8541	1	1	4	702
trauma	0.2589	0.4383	0	0	1	788
age	38.94	13.26	37	19	84	797

Factor❺

	1	2	3
f.type.centre	100	249	450
Percent	12.52	31.16	56.32

mode = 3   Valid n = 799

There are eight numerical variables ❶. Four are binary variables coded 1 (presence) and 0 (absence): scz.cons (consensus for a diagnosis of schizophrenia), dep.cons (consensus for depression), subst.cons (substance use disorder), and trauma (occurrence of trauma during childhood). Two variables are ordered: grav.cons (consensus on gravity, coded from 1 to 7) and char (character, the variable in the Cloninger instrument that determines the existence of a personality disorder) coded (0, 1, 2, 3) for "absent," "mild," "moderate," or "severe." These two ordered categorical variables will be now considered quantities. The last two numerical variables are age and dur.interv (duration of interview). There is one categorical variable❺: f.type.centre. It is obtained from the numerical variable type.centre2 using the instruction f.type.centre <- factor(type.centre2); it indicates whether a prisoner is detained in a high-security unit (level 1), a centre for longer sentences or for prisoners with good scope for rehabilitation (level 2), or a prison for remand prisoners or short sentences (level 3).

One of the main results is the number of missing data for each variable❷. Many variables have no or few missing data, but char (character) is available for only 702 prisoners among the 799. This is important because in a regression model, only subjects who have data for *all* the variables will be included in the analysis.

The mean of the outcome dur.interv is about an hour❸. Often, the columns "min" and "max" are very informative. At least one patient has a duration of 0 minutes❹; this is, of course, impossible and suggests that dur.interv should be examined closely:

❶
```
> table(mhp.mod$dur.interv[mhp.mod$dur.interv < 30])

 0   4  15  25
 1   1   1   5
```

We focus here on subjects with an interview duration shorter than 30 minutes❶. One has a duration of 4 minutes, one of 15 minutes, and five of 25 minutes. After discussion with the investigators, it is decided that durations of interview of 15 minutes or less are not acceptable (it means that either the value is incorrect or the evaluation is invalid). The three corresponding prisoners are thus discarded and a new data file named mhp.mod2 is created using the function subset() (see Section 9.2 for details about this function):

```
> mhp.mod2 <- subset(mhp.mod, dur.interv > 15)
> dim(mhp.mod)
[1] 799❶ 94
> dim(mhp.mod2)
[1] 746❷ 94
```

The sample size drops from 799❶ to 746❷ because the three patients with abnormal values have been discarded, along with all patients who had missing values for the variable dur.interv.

One of the conditions of validity of linear regression is that the residuals should have a normal distribution. This is often related to the normality of the outcome so that a density curve for dur.interv represented with a normal curve with the same mean and standard deviation is useful (Figure 5.1) (see Section 4.7 for an explanation of the script):

```
> m <- mean(mhp.mod2$dur.interv, na.rm = TRUE)
> stddev <- sd(mhp.mod2$dur.interv, na.rm = TRUE)
> rangenorm <- c(m - 3.72 * stddev, m + 3.72 * stddev)
```

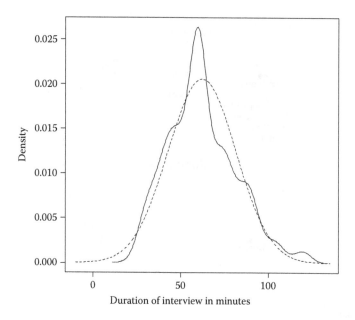

FIGURE 5.1

Density curve for the outcome variable "duration of interview" represented with a normal curve with same mean and standard deviation. Strictly speaking, the normality of the outcome is not a prerequisite for a linear regression; it is, nevertheless, a good habit to produce a diagram of this type (in case of a huge deviation from normality, another type of model could be considered).

```
> density1 <- density(mhp.mod2$dur.interv, na.rm = TRUE)
> density2 <- dnorm(pretty(rangenorm, 200), mean = m,
  sd = stddev)
> plot(density1, xlim = range(c(density1$x, rangenorm)),
  ylim = range(c(density1$y, density2)),
  xlab = "Duration of intervew", main = "")
> lines(pretty(rangenorm, 200), density2, lty = 2)
> box()
```

The distribution does not appear strictly normal, but the overall shape is not far enough from normality to stop the process at this point.

Before running the linear regression model, the bivariate associations between "duration of interview" on the one hand and each of the explanatory variables on the other should be considered. A series of t-tests or Pearson correlation tests could be appropriate for this. An interesting alternative is to use a focused principal component analysis (PCA) (Figure 5.2) (see Section 3.8). From the function fpca() in the package "psy" we obtain here:

```
> fpca(mhp.mod2, y = "dur.interv", x = c("scz.cons", "dep.cons",
  "subst.cons", "grav.cons", "char", "trauma", "age"), cx = 1)
```

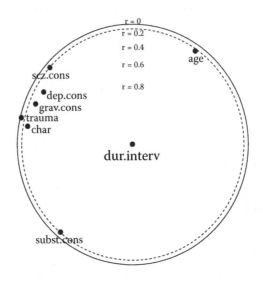

FIGURE 5.2
Focused principal component analysis of eight explanatory variables (the focus is on the outcome "duration of interview"). This representation provides the same information as a classic bivariate analysis. "Trauma during childhood" is the only variable not significantly associated with the outcome (it is outside the dotted circle). "Schizophrenia" and "substance abuse and dependence" are borderline. The diagram also suggests that "depression," "schizophrenia", "gravity", "trauma", and "character" are all correlated, while "substance abuse and dependence" and "age" are on their own.

We observe three clusters of variables containing "depression," "schizophrenia," "gravity," "traumatism," and "character" that seem to be correlated, while "substance abuse and dependence" and "age" are on their own. Because they are inside the dotted circle, "depression," "gravity," and "character" are significantly associated with "duration of interview." "Substance abuse and dependence" and "schizophrenia" are borderline and "trauma" is not significant.

Now it is time to run the linear regression model. The function lm() is dedicated to this:

```
        ❶              ❷
> mod <- lm(dur.interv ~ scz.cons + dep.cons + subst.cons
  + grav.cons + char+trauma + age + f.type.centre,
  data = mhp.mod2)
> summary(mod)❸

Call:
lm(formula = dur.interv ~ scz.cons + dep.cons + subst.cons
  + grav.cons + char + trauma + age + f.type.centre,
  data = mhp.mod2)
```

```
Residuals:
    Min        1Q      Median        3Q         Max
 -38.578    -13.855    -1.769     10.922      64.405

Coefficients:
                   Estimate   Std. Error❹  t value   Pr(>|t|)
(Intercept)        42.48996    3.99998      10.623   < 2e-16    ***
scz.cons❺           3.064206❻   2.80997       1.090   0.275911❼
dep.cons            6.71269    1.64793       4.073   5.21e-05   ***
subst.cons          4.60037    1.79854       2.558   0.010760   *
grav.cons           1.06236    0.56548       1.879   0.060737   .
char                1.62547    0.93536       1.738   0.082723   .
trauma             -0.66805    1.66931      -0.400   0.689144
age                 0.20788    0.06066       3.427   0.000649   ***
f.type.centre2❽     4.09540    2.53401       1.616   0.106546
f.type.centre3❾    -1.29681    2.44159      -0.531   0.595509
---
Signif. codes: 0 '***' 0.001 '**' 0.01 '*' 0.05 '.' 0.1 ' ' 1

Residual standard error: 18.02 on 645 degrees of freedom
(❿141 observations deleted due to missingness)
Multiple R-squared: 0.1086, Adjusted R-squared: 0.09619
F-statistic: 8.734 on 9 and 645 DF, p-value: 1.919e-12
```

The syntax❷ is as follows: to the left of the "~" is the outcome and to the right the predictors. The result of the analysis is stored in the object mod❶ because further computations must be performed on it. The function summary()❸ is first used to present the results. Concerning the variable "schizophrenia"❺, the coefficient is 3.06❻. This means that, on average and all other predictors being constant, it takes about three minutes more to conclude an interview with a prisoner who has a diagnosis of schizophrenia. This difference, however, is not statistically significant as $p > 0.05$❼. A noteworthy result is the number of subjects included in the analysis in ❿: About 154 prisoners among 799 were dropped due to missing values for at least one of the variables. It is also a good habit to look at the standard error of estimates ❹. When any one is "abnormally" large (it is unfortunately difficult to give a precise definition of "abnormally"), this can correspond to a model that is poorly designed. Most often, two explanatory variables are highly correlated, and it may be preferable to drop one from the model. In this situation there is "multicolinearity" of the two corresponding predictors: They are so close to each other that it has, in general, no meaning to look for the effect of one, with the other held constant. The focused PCA shown in Figure 5.2 can help detect these situations *a priori*: Two highly correlated explanatory variables will correspond to two superimposed points on the diagram.

In ❽ and ❾ there are two lines of results for the categorical variable "type of centre". This is consistent because this variable has three levels and it has therefore been converted to two binary variables—f.type.centre2 and f.type. centre3—f.type.centre2 = 1 if the prisoner belongs to a type 2 prison (longer sentences and/or prisoners with good scope for rehabilitation), and f.type.centre2 = 0 otherwise; f.type.centre3 = 1 if the prisoner belongs to a type 3 prison (remand prisoners and/or short sentences) and f.type. centre3 = 0 otherwise (see Section 6.2 for details concerning the coding of categorical variables in regression models). Hence, ❽ provides the comparison of type 2 prisons with type 1 prisons, and ❾ provides the comparison of type 3 prisons with type 1 prisons. This can, however, seem inconvenient, and a single p-value should be available to test the global effect of the variable "type of prison". This p-value can be obtained using the function drop1():

```
            ❶          ❷
> drop1(mod, .~., test = "F")
Single term deletions

Model:
dur.interv ~ scz.cons + dep.cons + subst.cons + grav.cons + char
    + trauma + age + f.type.centre

                Df  Sum of Sq   RSS   AIC  F value      Pr (F)
<none>                       209523  3798
scz.cons         1       386  209910  3797   1.1891  0.2759114
dep.cons         1      5390  214913  3813  16.5927  5.21e-05   ***
subst.cons       1      2125  211649  3803   6.5425  0.0107602   *
grav.cons        1      1147  210670  3800   3.5295  0.0607370   .
char             1       981  210504  3799   3.0199  0.0827234   .
trauma           1        52  209575  3796   0.1602  0.6891441
age              1      3815  213338  3808  11.7436  0.0006493   ***
f.type.centre   2❸     3446  212970  3805   5.3044  0.0051888❹**
---
Signif. codes: 0 '***' 0.001 '**' 0.01 '*' 0.05 '.' 0.1 ' ' 1
```

The instruction "~❶ asks that the global effect of all predictors be estimated. The instruction "test = "F""❷ is appropriate here because this is a linear model. The global effect of "type of centre" is in ❹; it is significant at the 5% level. It is interesting to note the degree of freedom in ❸, not necessarily from a theoretical point of view but simply because it is equal to the number of levels of the variable less 1: If f.type.centre had not been declared explicitly as a factor (with the function as.factor()), it would have been considered a numerical variable, and the number of degrees of freedom would have been equal to 1.

Now it is time to verify the conditions of validity of the model; first, the normality of residuals. The function plot() applied to the model gives a normal probability of residuals (Figure 5.3):

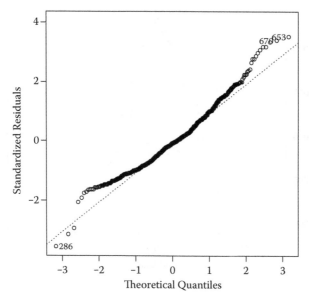

Im(dur.interv ~ scz.cons + dep.cons + subst.cons + grav.cons + char + trauma

FIGURE 5.3
Normal probability plot of residuals as obtained from "plot(model, 2)." At the extreme left and right, there are points above the diagonal, which suggests a skewed distribution.

```
> plot(mod, 2)
```

The distribution is approximately normal, but to the extreme left and right of the diagram there are points above the diagonal: This reflects a skewed distribution of residuals. A normal probability plot is an efficient system and is obtained directly from the model estimated using lm(). It is, however, not always easy to interpret. An alternative can be to draw a density curve of the residuals superimposed on the distribution of a normal variable with the same mean and standard deviation (Figure 5.4) (see Section 4.7 for an explanation of a similar script):

```
> m <- mean(resid(mod), na.rm = TRUE)
> stddev <- sd(resid(mod), na.rm = TRUE)
> rangenorm <- c(m - 3.72 * stddev, m + 3.72 * stddev)
> density1 <- density(resid(mod), na.rm = TRUE)
> density2 <- dnorm(pretty(rangenorm, 200), mean = m,
    sd = stddev)
> plot(density1, xlim = range(c(density1$x, rangenorm)),
    ylim = range(c(density1$y, density2)), xlab = "Residuals",
    main = "")
> lines(pretty(rangenorm, 200), density2, lty = 2)
> box()
```

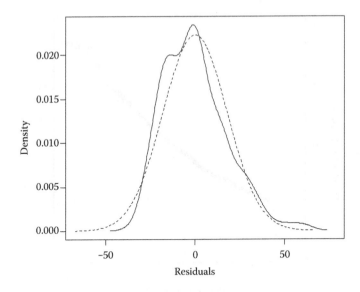

FIGURE 5.4
Density curve of residuals superimposed on a normal distribution with same mean and standard deviation. The top of the distribution of residuals is slightly inclined to the left; the distribution is "skewed." This diagram is likely less sensitive than a normal probability plot, but it is easier to interpret.

Obviously, the density of the residuals is skewed (leaning to the left), but only moderately, so that the results should be acceptable. A strict-minded statistician could, however, find this distribution unacceptable. To tackle this problem, several solutions can be proposed. The first is to convert the outcome (Y can be changed into the logarithm of Y, the square root of Y, or box-cox transformations can even be used (Fox 2002)). This can be statistically efficient but disappointing from a practical point of view: How can the effect of a covariate on a square root or on a logarithm of one minute be interpreted? The second solution is to use non-parametric approaches such as the bootstrap (see Section 6.7).

The second condition of validity of the linear regression model involves the variance of residuals, which should be constant. Constant here means that the residual is independent of Y and of the X_i values. This condition can appear somewhat academic: Why should the variance of residuals vary according to other variables? There are, in fact, very real situations where this is the case. For example, suppose the outcome Y is "expenditure on meals" and the explanatory variable X_1 is "income" (Wikipedia contributors 2009a); then a person with a low income will spend a fairly constant amount each day (not less because food has a minimum cost, and not more because the person cannot afford it), while a wealthier person may occasionally spend less (a sandwich) or occasionally more (an expensive restaurant). The variance of Y thus increases with X_1 so that the variance of residuals also

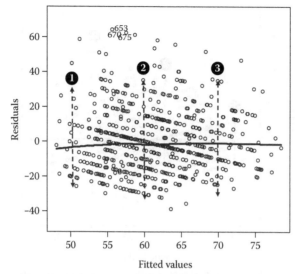

Im(dur.interv ~ scz.cons + dep.cons + subst.cons + grav.cons + char + trauma

FIGURE 5.5
Plot of residuals versus fitted values. The objective is to verify that the variance of residuals is constant and independent from the fitted values. This seems to be the case here, as the variations of residuals for 50 minutes❶, 60 minutes❷, and 70 minutes❸ are comparable.

increases with X_1. If it is perhaps disproportionate to verify that the variance of residuals is independent of Y and of all the X_i values, then it is classic, and a good habit, to ascertain graphically that the variance of residuals is independent of the "fitted values," that is, "$a_0 + a_1 \times X_1 + a_2 \times X_2 + ... + a_p \times X_p$." Furthermore, it is easy to obtain this graph in R (Figure 5.5):

```
> plot(mod, 1)
```

There are no obvious differences in residual variance across interview duration: For 50, 60, or 70 minutes, the residuals are roughly all between −30 and 30. A typical situation where the condition of constant variance is violated is presented in Figure 5.6. Indeed, variance obviously increases with time here. In this circumstance, two solutions can be proposed: (1) a conversion of Y or the X_i values (e.g., the regression is performed on log(Y) instead of Y), and (2) the use of a bootstrap procedure (Cribari-Neto and Zarkos 1999). The bootstrap is presented in Section 6.7.

The last condition of validity of the linear regression model is that the residual term ε must be independent of Y and the X_i values, and that ε has no pattern of autocorrelation. Regarding the independence of ε from Y and the X_i values, this is in fact a modelling problem. In a linear regression model, it is hypothesised that X_i is linearly related to Y (in our example, age and duration of interview). This is, of course, never entirely true. If the

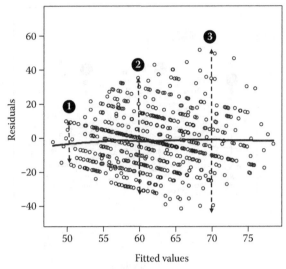

Fitted values

lm(dur.interv ~ scz.cons + dep.cons + subst.cons + grav.cons + char + trauma

FIGURE 5.6
Plot of residuals versus fitted values. This is a typical situation where residual variance increases with fitted values (variance for 50 minutes❶ is below the variance for 60 minutes❷, which is, in turn, below the variance for 70 minutes❸).

relationship is "approximately linear," then the model will be acceptable. If it is not, it will be necessary to consider a more sophisticated relationship such as $Y = \ldots + a_i \times \log(X_i) + \ldots$ or $Y = \ldots + a_i \times X_i + b_i \times X_i^2 + \ldots$. A good tool to investigate the linear nature of a relationship is the regression spline; it is presented in Section 6.1. Concerning a potential pattern of autocorrelation in the residual term ε, it can be inferred from the structure of the data. In our example, this pattern does exist. Indeed, if two prisoners live in the same prison, it is likely that their durations of interview will be closer than if they lived in different prisons. The explanatory variable f.type.centre addresses this issue, but not completely: Two prisons of the same type can have rather different habits, such that long interviews will be possible in one and not viewed favourably in the other. If a model does not address this issue, is it a major error? From a theoretical point of view, it is. From a real-life perspective, however, it is not necessarily so. Indeed, if the design effect is close to 1 (the design effect quantifies the level of similarity of detainees from the same prison), then the cost in terms of model complexity due to the inclusion of the "prison" factor can exceed the gain in terms of statistical orthodoxy. This point is discussed again in Section 6.8.

At this point, we must nevertheless acknowledge that the linear model obtained from lm() does not take into account the two-stage weighted sample design of the MHP study. Here again, this may or may not be a major problem. The point was discussed in Section 4.3 and will be discussed again in Section 6.8. In short, if the objective is to explain the "duration of interview" in French prisons for men in general, it is, in fact, necessary to take into account the particular sampling design, using the function svyglm() in the package "survey":

❶

```
> mod <- svyglm(dur.interv ~ scz.cons + dep.cons + subst.cons
                                               ❷
+ grav.cons + char + trauma + age, design = mhp.survey)
> summary(mod)

Call:
svyglm(dur.interv ~ scz.cons + dep.cons + subst.cons +
    grav.cons + char + trauma + age, design = mhp.survey)

Survey design:
svydesign(id = ~centre, strata = ~type.centre2, weights = ~pw,
    fpc = ~strat.size, data = mhp.mod2)

Coefficients:
```

	Estimate	Std. Error❸	t value	Pr(>|t|)	
(Intercept)	40.81049	6.63311	6.153	0.000108	***
scz.cons❹	4.13725❺	2.58003	1.604	0.139891❻	
dep.cons	6.24544	2.43400	2.566	0.028087	*
subst.cons	5.22647	2.56044	2.041	0.068505	.
grav.cons	0.80981	0.62176	1.302	0.221962	
char	2.17576	1.41826	1.534	0.156011	
trauma	-0.75968	1.13918	-0.667	0.519948	
age	0.26003	0.08254	3.150	0.010325	*

```
---
Signif. codes: 0 '***' 0.001 '**' 0.01 '*' 0.05 '.' 0.1 ' ' 1
(Dispersion parameter for gaussian family taken to be 343.8875)
Number of Fisher Scoring iterations: 2
```

The syntax in ❶ is comparable to the syntax we used with lm(). The object "mhp.survey"❷ includes data, weights, centres, and strata; it was constructed in Section 4.1 The results are presented in a similar way; the specific effect of having a diagnosis of schizophrenia❹ is now an increase of 4.14 minutes❺ as compared to 3.06 minutes in the previous model. The p-value is still above 5% in ❻. It can be noted that the standard deviations of the coefficients❸ are mostly larger than those obtained with lm(); this is generally the case, as explained in Section 4.1 and following. Finally, there are no longer any results concerning the variable f.type.centre because it is now directly included in the model through the instruction design =❷.

To conclude this discussion of linear regression models, a few words are required about "unusual data," sometimes also called "outliers." There are

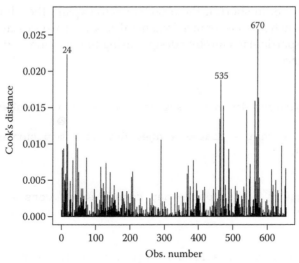

svyglm(dur.interv ~ scz.cons + dep.cons + subst.cons + grav.cons + char + trauma

FIGURE 5.7
Cook's distance for each observation in the dataset. This distance corresponds to a global estimate of the influence of a given prisoner on the estimations of the coefficients of the model. Here, observations 24, 535, and 670 are the most influential.

always unusual values in a dataset, and these data correspond in general to correct observations obtained for unusual but nevertheless real people. Basically, this is not a statistical problem. There are, however, two situations in which this can be a problem: (1) when data are unusual because they are false (e.g., arising from a problem of data capture), and (2) when unusual data are so numerous that this is incompatible with a residual with a normal distribution. Because one cannot be sure that these two situations did not occur in the analysis, some authors argue that it may be useful to explore the robustness of results concerning "unusual data." Here again, this can be done simply in R with the plot() function:

```
> plot(mod, 4)
```

Figure 5.7 presents Cook's distance for each observation. This distance is a numerical representation of the degree of modification of model coefficients when a given subject is dropped from the dataset. If the modification is large, the subject is considered "influential." Here, observations 24, 535, and 670 are the most influential. However, no observation has a Cook's distance that is substantially greater than the others, and no distance is, to any large degree, greater than 1. It is thus not necessary to pursue the investigation (Maindonald and Braun 2007). If such had been the case, it would have been necessary to verify that these observations did not correspond to impossible values or, if possible, that they did not correspond to errors in data capture. It may be also interesting to rerun the model without the main outliers in order to verify that the new results are basically similar. In this situation, the syntax in R is simply

```
> mod <- lm(dur.interv ~ scz.cons + dep.cons + subst.cons
   + grav.cons + char + trauma + age + f.type.centre,
                     ❶
   data = mhp.mod[-c(24, 535, 670), ])
> summary(mod)
```

The instruction -c[24, 535, 670]❶ discards the corresponding rows in the dataset mhp.mod.

5.2 Logistic Regression for Binary Outcome

In a few words: In many situations, the outcome Y is a binary variable (e.g., being single or not, unemployed or not, diabetic or not). If Y is coded 0 (no) and 1 (yes), it is theoretically possible to consider the linear regression model:

$$Y = a_0 + a_1 \times X_1 + a_2 \times X_2 + \ldots + a_p \times X_p + \varepsilon.$$

However, in such a situation, ε will have a distribution that is far from normal so that the statistical tests of hypothesis $a_i = 0$ will be unreliable. A solution to this problem consists of using $Log\left[\dfrac{prob(Y=1)}{1-prob(Y=1)}\right]$ instead of Y. This can appear rather mysterious, but the new term has the advantage of varying between $-\infty$ and $+\infty$ so that the new equation

$$Log\left[\frac{prob(Y=1)}{1-prob(Y=1)}\right] = a_0 + a_1 \times X_1 + a_2 \times X_2 + \ldots + a_p \times X_p$$

is more logical from a mathematical point of view. Furthermore, if X_1 is a binary predictor coded as 0 and 1 and if $X_2, \ldots, X_p$ are kept constant, then

$$Log\left(\frac{prob(Y=1)}{1-prob(Y=1)}\right)_{(X=1)} = a_1 + [a_0 + a_2 \times X_2 + \ldots + a_p \times X_p] \text{ and}$$

$$Log\left(\frac{prob(Y=1)}{1-prob(Y=1)}\right)_{(X=0)} = 0 + [a_0 + a_2 \times X_2 + \ldots + a_p \times X_p].$$

Then, by subtraction,

$$a_1 = Log\left(\frac{prob(Y=1)}{1-prob(Y=1)}\right)_{(X=1)} - Log\left(\frac{prob(Y=1)}{1-prob(Y=1)}\right)_{(X=0)}$$

or, equivalently,

$$e^{a_1} = \frac{\left(\dfrac{prob(Y=1)}{1-prob(Y=1)}\right)_{(X=1)}}{\left(\dfrac{prob(Y=1)}{1-prob(Y=1)}\right)_{(X=0)}}$$

The exponential of a_1 is thus equal to the odds-ratio associating the binary outcome Y to the binary predictor X_1, all other predictors being constant, and $exp(a_i)$ is thus known as an adjusted odds-ratio. When X_1 is not binary, $exp(a_1)$ is the adjusted odds-ratio associating Y to an increase of one unit of X_1. These fascinating results, combined with a certain robustness of the numerical procedure that estimates the a_i, explain the wide popularity of logistic regression.

It is, of course, possible to test the hypothesis $a_1 = 0$ (which can be expressed as follows: the predictor is associated with the outcome, all other predictors being constant). Like all statistical tests, this one has several conditions of validity. Unfortunately, it is difficult to set them out in a simple way.

A classic rule is that there should be at least five events per explanatory variable in the model (Vittinghoff and McCulloch 2007) or even 10 events per variable (Peduzzi et al. 1996). More clearly, if there are six explanatory X_i variables, there should be at least 30 (30 = 5 × 6) subjects for whom Y = 1 (and at least 30 subjects for whom Y = 0).

Another condition of validity that must be verified concerns the linear relationship that is postulated between $Log\left[\dfrac{prob(Y=1)}{1-prob(Y=1)}\right]$ and X_i.

A regression spline is a tool that can be used to deal with this point; this is detailed in Section 6.1.

Goodness-of-fit tests are sometime proposed. These tests set out to assess globally if the deviations observed between data and the logistic model are compatible with what could be observed by "chance." The advantage of these tests is that they are simple to use. Their disadvantage is that they do not really answer the right question. The need is to determine whether the deviations observed between the data and the model are small enough to obtain interpretable results, and not whether these deviations could be observed by "chance," which is the information provided by the p-value of a goodness-of-fit test.

Finally, as in the linear regression model, it will be useful to look for the potential influence on results of "unusual data."

In Practice: Let us imagine now that we are interested in explaining the binary outcome "being at a high risk for suicide attempt" by the following predictors: "abuse" (abuse occurring during childhood yes/no), "disciplinary procedure" (yes/no), "duration of sentence" (<1 month, 1--6 months, 6–12 month, 1–5 years, >5 years), "age" (continuous), and "type of prison" (categorical 1, 2, 3). All these explanatory variables except "type of prison" are considered numeric variables.

Here again, the description of all these variables is a basic but unavoidable step. Using the function describe() in the package "prettyR" we have

```
> describe(mhp.mod[, c("suicide.hr", "abuse",
    "discip", "duration", "age", "f.type.centre")],
    num.desc = c("mean", "sd", "median", "min", "max", "valid.n"))

Numeric
                  mean        sd    median    min    max   valid.n
suicide.hr      0.2013❶    0.4012      0         0      1     760❷
abuse           0.2765❸    0.4476      0         0      1     792
discip          0.2333❹    0.4232      0         0      1     793
duration        4.318      0.8521      5❺        1      5     575❻
age            38.94      13.26       37        19     84     797


Factor
                    1          2          3
f.type.centre     100        249        450
Percent         12.52❼     31.16❽     56.32❾
mode = 3   Valid n = 799
```

A major result is often found in the column relating to missing data. There are more than 200 missing observations for the variable "duration" (duration of sentence)❻. This could have been expected because remand prisoners, by definition, have no sentence at the moment of their interview. Concerning binary and categorical variables, it is preferable for all levels to have a sufficient number of observations (variables with levels counting as few as one or two events are likely to raise estimation problems). There is no difficulty of this nature here (❶, ❸, ❹, ❼, ❽, ❾). The number of "events" of the outcome variable is about $0.2❶ \times 760❷ \approx 150$, which is well above the minimum of 5 (or even 10) times the number of explanatory variables (here, 6). Finally, it is noteworthy that the median of "duration" is equal to its maximum❺. This is a well-known particularity of cross-sectional studies performed in institutions: Subjects staying a long time are over-represented.

The bivariate analysis can be performed at this stage, for example using a focused PCA as proposed in Section 5.1 for the linear regression model. It is not presented here.

The logistic regression model can finally be estimated using the glm() function:

```
                                    ❶
> mod <- glm(suicide.hr ~ abuse + discip + duration + age
  + f.type.centre, data = mhp.mod, family = "binomial"❷)
> summary(mod)

Call:
glm(formula = suicide.hr ~ abuse + discip + duration + age
  + f.type.centre, family = "binomial", data = mhp.mod)

Deviance Residuals:
   Min        1Q      Median        3Q        Max
-1.3079   -0.6922   -0.5746   -0.4724    2.1142

Coefficients:
                    Estimate  Std. Error❸ z value  Pr(>|z|)
(Intercept)       -0.329750    0.787283   -0.419   0.67533
abuse              0.633421❹   0.228495    2.772   0.00557❺ **
discip             0.454877    0.254880    1.785   0.07431  .
duration          -0.295252    0.149316   -1.977   0.04800  *
age               -0.005002    0.009468   -0.528   0.59729
f.type.centre2    -0.045704    0.333694   -0.137   0.89106
f.type.centre3     0.273555    0.371157    0.737   0.46110
---
Signif. codes:  0 '***' 0.001 '**' 0.01 '*' 0.05 '.' 0.1 ' ' 1

(Dispersion parameter for binomial family taken to be 1)

    Null deviance: 555.49 on 548 degrees of freedom
Residual deviance: 531.28 on 542 degrees of freedom
    (250 observations deleted due to missingness)
AIC: 545.28

Number of Fisher Scoring iterations: 4❻
```

The syntax ❶ is comparable to that used for the linear regression model. The instruction family = "binomial"❷ here specifies that a logistic model is required. The results are also similar to the output of lm(). The coefficient a_i relating to the variable "abuse" is in ❹, the p-value of the test $a_i = 0$ is in ❺ and is significant at the 5% level: Abuse during childhood is a risk factor for high risk of suicide attempt (controlling for the effects of "duration of sentence," "disciplinary procedure," "age," and "type of prison"). The variable "duration of sentence" is negatively associated with the outcome (the shorter the sentence, the greater the risk of suicide attempt). The other variables are not significantly associated with the outcome.

As in linear regression, it is a good habit to check that there is no major problem of multicolinearity in ❸ (no standard error is "abnormally" great). The number of iterations required by the algorithm that determines the maximum likelihood is in ❻. This number depends on the complexity and on the shape of the likelihood function around its maximum. Here, 4 is a good value. In general, it is all the higher when there are a large number of explanatory variables. When it is greater than 25, R stops and issues an error message.

The coefficients provided in the output are a_i coefficients and not $\exp(a_i)$, the adjusted odds-ratio. The functions `exp()`, `coefficients()`, and `confint()` can be used to estimate the adjusted odds-ratios and their confidence intervals:

```
> exp(coefficients(mod))
   (Intercept)         abuse❶       discip     duration       age
    0.7191038       1.8840458    1.5759788   0.7443441  0.9950108
f.type.centre2 f.type.centre3
    0.9553243       1.3146295❷
```

```
> exp(confint(mod))
Waiting for profiling to be done...
                     2.5%        97.5%
(Intercept)      0.1526791    3.367324
abuse            1.1996784    2.943396❸
discip           0.9514471    2.590155
duration         0.5549039    0.997894
age              0.9764391    1.013444
f.type.centre2   0.5042208    1.878234
f.type.centre3   0.6406323    2.763229
```

The adjusted odds-ratio associated with "abuse" is 1.88❶, with a 95% confidence interval of [1.20,2.94].

Because `f.type.centre` is a three-levels categorical variable, it requires two a_i coefficients (see Section 6.2 for more details concerning this point). In ❷, the odds-ratio between the outcome (high risk of suicide attempt) and level 3 prisons (remand prisoners or short sentences) compared to level 1 prisons (high-security prison) in the variable "type of centre" is 1.31. There is, however, no single answer to the following question: Is there a "type of centre" effect on high risk of suicide attempt? As in the case of linear regression modelling, the function `drop1()` can be used for this purpose:

```
                              ❶
> drop1(mod, .~., test = "Chisq")
Single term deletions

Model:
suicide.hr ~ abuse + discip + duration + age + f.type.centre
```

```
                Df    Deviance   AIC    LRT    Pr(Chi)
<none>         531.28  545.28
abuse           1      538.78  550.78  7.50  0.006166   **
discip          1      534.41  546.41  3.13  0.076924   .
duration        1      535.18  547.18  3.90  0.048379   *
age             1      531.56  543.56  0.28  0.596049
f.type.centre   2      532.72  542.72  1.44  0.487522❷
---
Signif. codes:  0 '***' 0.001 '**' 0.01 '*' 0.05 '.' 0.1 ' ' 1
```

The instruction test = "Chisq"❶ is used here because it is a logistic regression. The global p-value of the variable f.type.centre is in ❷; it is greater than 0.05: There is no significant effect of the type of prison on the risk of suicide attempt (controlling for the other predictors).

The package "epicalc" has a function logistic.display() that provides an appealing and synthetic presentation of logistic regression results obtained from glm() (however, some caution is required concerning the syntax used in the glm model; see the R documentation of logistic.display() for more details):

```
> library(epicalc)
> logistic.display(mod❶, decimal = 3)

Logistic regression predicting suicide.hr
                       ❷                  ❸           ❹         ❺
                                                    P(Wald's     P
                   crude OR(95%CI)    adj. OR(95%CI)  test)   (LR-test)
abuse: 1 vs 0  1.878 (1.214,2.903) 1.884 (1.204,2.948)  0.0056  0.0062
discip: 1 vs 0 1.617 (1.026,2.55)  1.576 (0.956,2.597)  0.0743  0.0769
duration       0.693 (0.551,0.871) 0.744 (0.555,0.997)  0.048   0.0484
  (cont.var.)
age (cont.var) 0.980 (0.964,0.996) 0.995 (0.976,1.013)  0.5973  0.596
f.type.centre: ref.=1                                            0.4875
    2          1.047 (0.557,1.968) 0.955 (0.497,1.837)  0.8911
    3          1.787 (0.962,3.32)  1.315 (0.635,2.721)  0.4611

Log-likelihood = -265.64098
No. of observations = 549
AIC value = 545.28196
```

The object mod❶ was estimated earlier using the glm() function. Along the same line, the following are presented: the "crude" odds-ratio❷ (i.e., the bivariate odds-ratio as computed in Section 3.1), the adjusted odds-ratio❸ obtained from the logistic regression model, Wald's test❹ (provided by glm()), and the likelihood-ratio test❺ (provided by drop1()). It can be noted that the confidence intervals in ❸ are slightly different from those obtained above from `exp(confint(mod))`. This is because the two functions use two different estimation procedures. The one used by `exp(confint(mod))` is preferable from a theoretical point of view (Faraway 2006).

Concerning the conditions of validity, we have already seen that the "number of events per explanatory variable" was acceptable (it should be at least equal to 5 or 10; it is greater than 20 here). Concerning the relevance of the hypotheses of linearity, it can be assessed using regression splines, as discussed in more detail in Section 6.1.

Concerning goodness-of-fit tests, we saw above that the value of these tests is not always obvious: if one is working with a sample of one million subjects, the fit is likely to be poor whatever the quality of the model (with such a large volume of information, it is as if the model is under a microscope, so that all the imperfections become obvious). Conversely, with a sample of about 50 subjects, the fit is, in general, likely to be good. Nevertheless, these tests can "give an idea" about the model fit, and they are indeed sometimes required by experts. The functions `lrm()` and `resid()` in the "Design" package can be used to compute the Hosmer and Le Cessie goodness-of-fit test (Hosmer et al. 1997):

```
> library(Design)
  ❶
> mod2 <- lrm(suicide.hr ~ abuse + discip + duration + age
  + factor(type.centre2), x = TRUE❷, y = TRUE❷, data = mhp.mod)
> resid(mod2, 'gof')
```

Sum of squared errors	Expected value|H0	SD
85.1012708	85.0174639	0.2573651
Z	P	
0.3256344	0.7447010❸	

The syntax of `lrm()`❶ is close to the syntax of glm(). The instructions ❷ are necessary to obtain the goodness-of-fit test later in the process. The p-value of this test is in ❸. It is greater than 0.05 so that the hypothesis of a good fit cannot be rejected (and, by default, it is thus accepted).

Concerning the possible influence of "unusual data," the point was addressed in Section 5.1 concerning linear regression models; the principle is comparable here:

```
> plot(mod, 4)
```

No distance is substantially greater than the others and clearly, none is greater than 1 (Figure 5.8). The question of possible influential observations is thus not a real problem here.

None of these analyses take into account the "prison" effect (only the "type of prison" effect). Neither do they take into account the two-level weighted

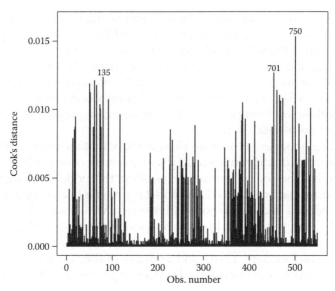

glm(suicide.hr ~ abuse + discip + duration + age + f.type.center)

FIGURE 5.8
Cook's distances for a logistic regression model. No observation registers a distance substantially greater than the others, and no distance is greater than 1. The problem of possible
influential observations is thus not crucial here.

sampling design of the MHP study. The validity of previous results is thus
an issue (discussed in Section 5.1). If the objective is to make inferences concerning the whole population of French prisoners, the svyglm() function in
the "survey" package can be used:

```
                              ❶
> mod <- svyglm(suicide.hr ~ abuse + discip + duration + age,
                  ❷
    design = mhp.survey, family = "binomial")
> summary(mod)

Call:
svyglm(suicide.hr ~ abuse + discip + duration + age,
    design = mhp.survey, family = "binomial")

Survey design:
svydesign(id = ~centre, strata = ~type.centre2, weights = ~pw,
    fpc = ~strat.size, data = mhp.mod)

Coefficients:
              Estimate    Std. Error   t value   Pr(>|t|)
(Intercept)   -0.048479    0.648534     -0.075    0.9416
abuse          0.697983❸   0.245036      2.848    0.0137❹*
```

```
discip          0.428479      0.218196      1.964      0.0713      .
duration       -0.353822      0.124543     -2.841      0.0139      *
age            -0.002942      0.009146     -0.322      0.7528
---
Signif. codes:  0 '***' 0.001 '**' 0.01 '*' 0.05 '.' 0.1 ' ' 1

(Dispersion parameter for binomial family taken to be 0.8995884)

Number of Fisher Scoring iterations: 4

> exp(coefficients(mod))
(Intercept)      abuse         discip      duration        age
0.9526777      2.0096951❺    1.5349206   0.7019997    0.9970625
> exp(confint(mod))
                   2.5%          97.5%
(Intercept)    0.2672472      3.3960867
abuse          1.2432404      3.2486673❻
discip         1.0008234      2.3540427
duration       0.5499531      0.8960829
age            0.9793485      1.0150969

Warning message:
In eval(expr, envir, enclos) : non-integer #successes in a
   binomial glm!
```

The syntax in ❶ is similar to the syntax of glm(). The object "mhp.survey"❷ was constructed from the original dataset in Section 3.1. Concerning the variable "abuse," the coefficient a_i is in ❸, the corresponding p-value in ❹, the adjusted odds-ratio in ❺, and its confidence interval in ❻.

5.3 Logistic Regression for a Categorical Outcome with More than Two Levels

In a few words: Let us consider a marketing study on cultural habits. Three groups of subjects are considered:

1. Those who prefer buying books on the Internet,
2. Those who prefer buying books in traditional bookshops, and
3. Those who buy less than one book a year.

The objective is to compare the characteristics of group 1 versus group 3 and those of group 2 versus group 3. The outcome is a categorical variable with three levels, and we have no tools at the moment to model it in a regression. However, because the focus is on the comparisons of groups 1 and 2 with group 3, two logistic regressions can be implemented: one that will discard group 2 (the outcome will thus be "buy books on the Internet" or "do not buy books") and one that will discard group 3 (and the outcome will be "buy books in bookshops" or "do not buy books"). This approach is definitely possible; it has, however, at least one drawback: it is not possible to statistically compare the strength of association of a given predictor with each outcome, as estimated in the two models. For example, "age" will possibly be more strongly associated with "buying books in bookshops" than with "buying books on the Internet." If two separate logistic regressions are used, it will not be possible to statistically compare the two odds-ratios because the control group is common to both analyses. To deal with this feature, a new type of model must be used: multinomial models.

Before engaging in a multinomial model, the first step is a series of logistic regressions that will contrast each level of the outcome with the reference level. This phase is primarily designed to verify the conditions of validity. The multinomial model will be estimated in the second stage. It will provide more consistent estimates and will enable the comparison of coefficients.

In Practice: In psychiatry, depression is a dimensional disorder. There can be mild, moderate, or severe depressive disorders. In the MHP study, it is possible to ask: Are the risk factors for mild or moderate depression comparable to the risk factors for severe depression? It is possible to operationalise this question in the following way. The outcome will be obtained from the variable "consensus for depression" and "consensus for gravity." More precisely, if a patient has no depression, the score will be 0; if the patient has a diagnosis of depression but a severity score under 5 ("notably ill"), the score will be 1; if there is depression with a severity score of 5 or more, then the score will be 2. The predictors selected are similar to those included in the model that explained high risk of suicide attempt in the previous section: "childhood abuse" (yes/no), "disciplinary procedure" (yes/no), "sentence duration" (<1 month, 1–6 months, 6–12 month, 1–5 years, >5 years), "age" (continuous), and "type of prison" (categorical 1, 2, 3). We are particularly interested here in a multinomial regression model because we hypothesise that the explanatory variable "abuse" will be more strongly associated with "severe depression" than with "non-severe depression."

As mentioned above, the first stage of the analysis should be to describe the variables, to carry out the bivariate analysis, and to estimate the two logistic regression models: "no depression" versus "non-severe depression" and "no depression" versus "severe depression." We now consider that all these analyses are complete and that the corresponding conditions of validity have been verified.

The function multinom() is called from the package "nnet":

```
> library(nnet)
> options(digits = 3)❶
                          ❷
> mod1 <- multinom(dep.ord ~ abuse + discip + duration + age
  + f.type.centre, data = mhp.mod)
# weights: 24 (14 variable)
initial value 622.913168
iter 10 value 511.048166
iter 20 value 509.266336
iter 20 value 509.266336
iter 20 value 509.266336
final value 509.266336
converged

> summary(mod1)

Call:
multinom(formula = dep.ord ~ abuse + discip + duration + age
  + f.type.centre, data = mhp.mod)
```

Coefficients:❸

	(Intercept)	abuse	discip	duration	age	f.type. centre2	f.type. centre3
1	0.796	-0.318❹	-0.209	-0.384	-0.00149	-0.402	-0.3752
2	0.360	0.463❺	0.454	-0.302	-0.00273	-0.547	-0.0327

Std. Errors:❻

	(Intercept)	abuse	discip	duration	age	f.type. centre2	f.type. centre3
1	0.880	0.288	0.313	0.167	0.00985	0.346	0.400
2	0.777	0.224	0.248	0.148	0.00922	0.311	0.341

```
Residual Deviance: 1019
AIC: 1047
```

The number of significant digits in the output is fixed in ❶. The syntax in ❷ is close to the syntax used in glm(). The coefficients are in ❸. It is surprising to note that the effect of "childhood abuse" is negative❹ when "not depressed" prisoners are compared to "non-severe depression" prisoners (negative means that the variable protects). It is the reverse when "non- depressed" prisoners are compared to prisoners with "severe depression"❺. But no conclusion can be drawn at this point because there are no p-values, only the standard deviations of each coefficient in ❻.

We now require odds-ratios with confidence intervals instead of coefficients with standard deviations. The functions coef() and confint() can be used for this purpose:

```
> exp(coef(mod1))
```

	(Intercept)	abuse	discip	duration	age	f.type.centre2	f.type.centre3
1	2.22	0.728❶	0.811	0.681	0.999	0.669	0.687
2	1.43	1.589❷	1.574	0.739	0.997	0.579	0.968

```
> exp(confint(mod1))
, , 1
```

	2.5%	97.5%
(Intercept)	0.395	12.433
abuse	0.414	1.280❸
discip	0.439	1.498
duration	0.491	0.945
age	0.979	1.018
f.type.centre2	0.340	1.319
f.type.centre3	0.314	1.505

```
, , 2
```

	2.5%	97.5%
(Intercept)	0.312	6.571
abuse	1.025	2.462❹
discip	0.968	2.559
duration	0.553	0.989
age	0.979	1.015
f.type.centre2	0.315	1.064
f.type.centre3	0.496	1.887

Concerning the "childhood abuse" risk factor, the 95% confidence interval❸ of the odds-ratio❶ in the first model ("non-severe depression" versus "no depression") contains 1. The association is thus not statistically significant at the 5% level. Conversely, the 95% confidence interval❹ of the odds-ratio❷ in the second model does not contain 1, so that "childhood abuse" appears to be a statistically significant risk factor for the "severe depression" outcome.

Even if the odds-ratio❷ is significantly different from 1 and not the odds-ratio❶, it is not possible to say at this point that ❷ is significantly greater than ❶. If we want to test this, we need to consider the difference between the corresponding coefficients in the output of multinom() (in ❹ and ❺ in the previous output). This difference is noted: $d = abuse(1) - abuse(2)$. If we had an estimate of variance of d, it would be possible to test "$d = 0$" from the statistic $d/var(d)^{1/2}$ which, under the null hypothesis, has a normal distribution with zero mean and unit variance. However,

```
var(d) = var(abuse(1) - abuse(2))
       = var(abuse(1)) - 2 × cov(abuse(1) - abuse(2)) + var(abuse(2))
```

From the function vcov() it is possible to estimate this quantity. Then $d/var(d)^{1/2}$ and finally the corresponding p-value:

```
> vdiff <- vcov(mod1)["1:abuse", "1:abuse"] - 2 * vcov(mod1)
  ["2:abuse", "1:abuse"] + vcov(mod1)["2:abuse", "2:abuse"]
> z <- (coefficients(mod1)[1, "abuse"] - coefficients(mod1)
  [2, "abuse"])/sqrt(vdiff)
> 2 * (1 - pnorm(abs(z)))
[1] 0.01456495❶
```

Because $p < 0.05$❶, it is now possible to conclude that the association between "childhood abuse" and "depression" is stronger for the phenotype "severe depression" than for the "non-severe depression" phenotype.

5.4 Logistic Regression for an Ordered Outcome

In a few words: Sometimes the outcome has more than two levels, and in this case they are ordered levels. For instance, in the previous example, "non-severe depression" was between "no depression" and "severe depression." Although this point was neglected in the analysis, valuable results were nevertheless obtained. In some circumstances, an option of this nature can result in a lack of statistical power and a model that is not sufficiently parsimonious. Another option takes into account the ordered nature of the outcome; it is known as the "proportional odds model." It is hypothesised here that there is an overall progression from "no depression" (0) to "non-severe depression" (1) and "severe depression" (2). Apart from this overall progression, the effect of each predictor is similar when [level (0) is grouped with level (1)] and compared to level (2), or when level (0) is compared to [level (1) grouped with (2)]. The model thus estimates a common odds-ratio for each predictor: This is more parsimonious, potentially easier to interpret, and more powerful from a statistical point of view.

To be confident in the "proportional odds model," the two logistic regressions should first of all be run: [levels (0) and (1)] versus level (2) and level (0) versus [levels (1) and (2)]. The condition of validity of these models should be verified carefully and the similarity between the two series of coefficients examined qualitatively. If everything is as it should be, the proportional odds model can be estimated. A statistical test is available to assess whether its fit is significantly different from the fit of the multinomial model, which is more general and not dependent on the proportional odds hypothesis.

In Practice: If we continue with the same example, we now have to estimate two logistic regressions and to compare the corresponding regression
coefficients. The first will study the effect of the predictors "childhood abuse,"
"disciplinary procedure," "sentence duration," "age," and "type of prison"
on the outcome "depression" ("no" versus "yes, severe or not"). The second
logistic regression will study the effect of the same predictors on the outcome
"depression" ("no or yes but non-severe" versus "yes, severe").

We obtain the following:

```
> mhp.mod$dep.0vs12 <- ifelse(mhp.mod$dep.ord == 0, 0, 1)❶
> mhp.mod$dep.01vs2 <- ifelse(mhp.mod$dep.ord == 2, 1, 0)❷
> table(mhp.mod$dep.ord, mhp.mod$dep.0vs12)❸

    0   1

 0  482   0
 1  0    131
 2  0    182


> table(mhp.mod$dep.ord, mhp.mod$dep.01vs2)❸

    0   1
 0  482   0
 1  131   0
 2  0    182


> coefficients(glm(dep.0vs12 ~ abuse + discip + duration + age
   + f.type.centre, data = mhp.mod, family = "binomial"))
    (Intercept)        abuse        discip    duration        age
       1.22039       0.17011❹    0.20590❺  -0.33459❻  -0.00212❼
f.type.centre2 f.type.centre3
    -0.48195❽       -0.16844❾


> coefficients(glm(dep.01vs2 ~ abuse + discip + duration + age
   + f.type.centre, data = mhp.mod, family = "binomial"))
    (Intercept)        abuse        discip    duration        age
      -0.37616       0.52648❹    0.49749❺  -0.21020❻  -0.00248❼
f.type.centre2 f.type.centre3
    -0.46519❽       0.04431❾
```

In ❶ and ❷, the two new outcomes have been constructed: depression "0"
versus ["1" & 2"] and depression ["0" & "1"] versus "2" (if dep.ord is equal
to 0, then dep.0vs12 is set to 0, otherwise it is set to 1❶). In ❸, two tables are
presented to verify that the coding conforms. Then two series of coefficients
appear. There is a certain discrepancy for "childhood abuse"❹ (0.17 versus
0.53) and "disciplinary procedure"❺ (0.21 versus 0.50). For the other variables,

there is good similarity (❻, ❼, ❽, and ❾). It is now possible to embark on the proportional odds model:

```
> library(MASS)
                         ❶
> mod2 <- polr(dep.ord ~ abuse + discip + duration + age
  + f.type.centre, data = mhp.mod)
> summary(mod2)

Re-fitting to get Hessian

Call:
polr(formula = dep.ord ~ abuse + discip + duration + age
  + f.type.centre, data = mhp.mod)

Coefficients:
                    Value    Std. Error   t value
abuse             0.2864❹    0.18791       1.524
discip            0.2905❺    0.20813       1.396
duration         -0.3006❻    0.12045      -2.496
age              -0.0021❼    0.00727      -0.289
f.type.centre2   -0.4752❽    0.24796      -1.916
f.type.centre3   -0.1050❾    0.27742      -0.379

Intercepts:
      Value   Std. Error   t value
0|1  -0.994    0.630        -1.576
1|2  -0.206    0.629        -0.327

Residual Deviance: 1028.75
AIC: 1044.75
(232 observations deleted due to missingness)
```

The syntax ❶ is classic. The coefficients are in ❹, ❺, ❻, ❼, ❽, and ❾. As foreseen, they are between the two corresponding series that were estimated above. It is possible to estimate the corresponding odds-ratios with their 95% confidence intervals using the instructions coef() and confint():

```
> exp(coef(mod2))
       abuse       discip   duration     age    f.type.centre2
       1.332        1.337      0.740     0.998        0.622
f.type.centre3
       0.900

> exp(confint(mod2))
Waiting for profiling to be done...
```

```
Re-fitting to get Hessian
```

	2.5%	97.5%
abuse	0.919	1.922
discip	0.887	2.008
duration	0.584	0.937
age	0.984	1.012
f.type.centre2	0.383	1.015
f.type.centre3	0.523	1.555

Model fit and conditions of validity are mainly derived from the two logistic regression models that were estimated first. It is, however, possible to test that this model fit is not significantly different from the fit of the multinomial model (which is more general). This test can be criticized, as its result depends a lot on the sample size, but it can give a rough idea of model acceptability. In practice, the deviances of the two models are subtracted and compared to a chi-square distribution with a degree of freedom equal to the difference in the number of parameters in the two models:

```
> pchisq(deviance(mod2) - deviance(mod1), mod1$edf - mod2$edf,
    lower.tail = FALSE)
[1] 0.1156693❶
```

The p-value is above 0.05, and the proportional odds model has a fit that is not significantly different from the multinomial model.

5.5 Regression Models for an Outcome Resulting from a Count

In a few words: In many questionnaire studies, the outcome Y results from a count: for example, number of cars or computers owned by a household, number of past suicide attempts, etc. The particularity of such outcomes is that they are quantities that follow a distribution that is far from normal (in the last example, there are many "0"s, some "1"s, and only a few "2 or more"s; the distribution is "L"-shaped). Linear regression is thus inapplicable, and an alternative must be found. Classically, the solution is to model log(Y) instead of Y:

$$\log(Y + \varepsilon) = a_0 + a_1 \times X_1 + a_2 \times X_2 + \ldots + a_p \times X_p$$

As in linear regression, ε is a "noise" with mean 0. But in the present situation it does not follow a normal distribution, in particular because the logarithm function requires a strictly positive argument. The a_i can be estimated and tested if it is hypothesised that Y follows a "Poisson distribution."

In other words, the probability that the numerical outcome Y is equal to k (k = 0, 1, 2, ...) must be

$$prob(Y = k) = \frac{\lambda^k e^{-\lambda}}{k!},$$

where λ is the mean of Y. This relationship can appear artificial. It is, however, encountered in some real situations, from the number of soldiers killed by horse-kicks each year in each corps in the Prussian cavalry to the number of phone calls at a call centre per minute (Wikipedia contributors 2009b). Unfortunately, the Poisson distribution has very strict prerequisites, notably that the mean of Y be equal to its variance. In practice, this can be far from true (Berk and MacDonald 2008), so that a Poisson regression is interesting mainly from a theoretical point of view.

Fortunately, these problems have been overcome using Quasi-Poisson or negative binomial regressions. These models are fairly similar to Poisson regression modelling but they are more robust, in particular concerning the question of equality of mean and variance (Gardner, Mulvey, and Shaw 1995; Ver Hoef and Boveng 2007). The advantage of Quasi-Poisson regression is that it is closer to the original Poisson model; most users will thus remain on familiar ground. The advantage of negative binomial models is that they are associated with a formal likelihood, which makes it easier to compare several models (Zeileis, Kleiber, and Jackman 2008).

There is, however, a rather frequent situation in which these last two models are not adequate: the so-called "excess zero" situation. For instance, if Y corresponds to the "number of past suicide attempts"variable, there is a difference between "0" on the one hand, and "1, 2, or more" on the other. One can indeed imagine that there is a barrier for attempting suicide and that if this barrier has been crossed once, there is a greater risk of recurrence. Formally, this will correspond to a distribution of suicide attempt with an excess of zero compared to a Poisson distribution. Certain models have been developed to tackle this problem, among which is the "hurdle model" (Zeileis, Kleiber, and Jackman 2008).

Concerning the conditions of validity of these models, there is no clear-cut state-of-the-art approach. Most of the conditions presented in the logistic regression section can also be used here: The linear relationship that is postulated between log(Y) and X_i must be at least approximately correct (the regression spline will be useful here; see Section 6.1); it is also interesting to look for the potential influence of "unusual data" on results.

In Practice: Let us imagine now that we are interested in explaining the outcome "number of past imprisonments" in the MHP study. This outcome clearly results from a count, and there is possibly an "excess zero" situation because prison is supposed to prevent the occurrence of repeat offences. The explanatory variables could be (1) "character," the Cloninger variable that

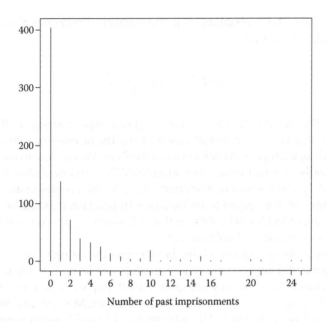

FIGURE 5.9
Barplot representing the distribution of the variable "number of past imprisonments." This distribution has an "L" shape, far from the bell-curve aspect that would be necessary to run a linear regression model. The large number of "0"s could correspond to a "zero excess" phenomenon.

determines the existence of a personality disorder, coded 0, 1, 2, 3 for absent, mild, moderate, severe; (2) "novelty seeking," another Cloninger temperament variable that is coded 0, 1, 2 for low, average, high; (3) family history of imprisonment, coded 0, 1 for no, yes; (4) "childhood abuse," coded 0, 1 for no, yes; (5) "age," continuous; and (6) "type of prison" (categorical, 1, 2, 3).

The first part of the analysis is descriptive. To limit the outputs, we focus here on the outcome. First, a barplot (Figure 5.9) to represent its distribution:

```
> plot(table(mhp.mod$nb.prison),
   xlab = "Number of past imprisonments", ylab = "")
```

The distribution is typically "L"-shaped. Let us now see the mean and variance of Y:

```
> mean(mhp.mod$nb.prison, na.rm = TRUE)
[1] 1.839949
```

```
> var(mhp.mod$nb.prison, na.rm = TRUE)
[1] 11.94743
```

The difference is manifest: This phenomenon is called "over-dispersion." It can occur for many reasons, for example the non-independence of the events that are counted. In the present situation, it can be hypothesised that once a prisoner has been in prison several times, the probability of a new imprisonment increases. The Poisson regression model is thus unusable. To deal with over-dispersion, a Quasi-Poisson regression model can be used with the glm() function and the quasipoisson option:

❶
```
> mod <- glm(nb.prison ~ char + ns + family.prison + abuse + age
   + f.type.centre, data = mhp.mod, family = "quasipoisson")

> summary(mod)

Call:
glm(formula = nb.prison ~ char + ns + family.prison + abuse
   + age + f.type.centre, family = "quasipoisson", data = mhp.mod)

Deviance Residuals:
   Min        1Q      Median       3Q        Max
-3.9882   -1.6538   -1.2669    0.2899    9.0217
```

Coefficients:❷
```
                   Estimate  Std. Error  t value  Pr(>|t|)
(Intercept)       -1.232314   0.418080    -2.948  0.003322 **
char               0.300950   0.069140     4.353  1.57e-05 ***
ns                 0.236566   0.088526     2.672  0.007730 **
family.prison      0.536230   0.140820     3.808  0.000154 ***
abuse              0.402111   0.139148     2.890  0.003988 **
age                0.006819   0.005590     1.220  0.223032
f.type.centre2     0.322015   0.261338     1.232  0.218345
f.type.centre3     0.276732   0.255984     1.081  0.280090
---
Signif. codes:  0 '***' 0.001 '**' 0.01 '*' 0.05 '.' 0.1 ' ' 1

(Dispersion parameter for quasipoisson family taken to be
   5.511868❸)

   Null deviance: 2834.4 on 633 degrees of freedom
Residual deviance: 2360.4 on 626 degrees of freedom
   (165 observations deleted due to missingness)
AIC: NA

Number of Fisher Scoring iterations: 6❹
```

The syntax ❶ and the results ❷ are similar to the output of a linear or logistic regression model. The dispersion parameter❸ would have been equal to 1 under the assumption of a Poisson distribution of Y. This is clearly not the case here. As in a logistic regression model, it is a good habit to verify that the maximisation process of the likelihood was smooth, and this is the case in ❹ (only six iterations).

The exponential of the coefficients can be useful. When the predictor X_i is binary, $\exp(a_i)$ is the number by which the numerical outcome is to be multiplied when passing from $X_i = 0$ to $X_i = 1$, all other predictors being constant. The functions coef() and confint() can be used here:

```
> exp(coef(mod))
(Intercept)         char              ns      family.prison      abuse
  0.2916171    1.3511419❶    1.2668914         1.7095494    1.4949771❷
                  f.type.         f.type.
       age        centre2         centre3
  1.0068421    1.3799049       1.3188132
```

```
> exp(confint(mod))
Waiting for profiling to be done...
                      2.5%          97.5%
(Intercept)      0.1264749      0.6527475
char             1.1781075      1.5452708
ns               1.0670383      1.5103039
family.prison    1.2963707      2.2528349
abuse            1.1359377      1.9612615
age              0.9957032      1.0177779
f.type.centre2   0.8474484      2.3755254
f.type.centre3   0.8202397      2.2507389
```

An increase of one slot for the variable "character" (e.g., from absent to mild or mild to moderate) shows an average number of past imprisonments multiplied by 1.35❶. A history of childhood abuse shows an average number of past imprisonments increased by about 50%❷.

The global effect of the variable "type of centre" can be obtained using the function drop1() and the search for influential data with the instruction plot(mod,4), with the same interpretations as in a linear or a logistic regression.

Of course, these analyses do not take into account the two-stage sampling design and the unequal weights in the MHP study. Results should be considered cautiously (see Section 5.1). In particular, a generalisation of results to the population of French male prisoners is not possible. If this generalisation is of interest, the function svyglm() in the package survey can be used:

```
                            ❶
> mod <- svyglm(nb.prison ~ char + ns + family.prison + abuse
                                    ❷
    + age + f.type.centre, data = mhp.mod, design = mhp.survey,
    family = "quasipoisson")
> summary(mod)
```

```
Call:
svyglm(nb.prison ~ char + ns + family.prison + abuse + age
    + f.type.centre, data = mhp.mod, design = mhp.survey,
    family = "quasipoisson")
```

```
Survey design:
svydesign(id = ~centre, strata = ~type.centre2, weights = ~pw,
    fpc = ~strat.size, data = mhp.mod)
```

Coefficients:❸

	Estimate	Std. Error	t value	Pr(>|t|)	
(Intercept)	-1.367320	0.358934	-3.809	0.00343	**
char	0.284851	0.085967	3.313	0.00783	**
ns	0.240724	0.058529	4.113	0.00210	**
family.prison	0.553971	0.142198	3.896	0.00298	**
abuse	0.381914	0.125414	3.045	0.01235	*
age	0.010622	0.005072	2.094	0.06266	.
f.type.centre2	0.315500	0.196977	1.602	0.14030	
f.type.centre3	0.300225	0.139505	2.152	0.05686	.

```
---
Signif. codes:  0 '***' 0.001 '**' 0.01 '*' 0.05 '.' 0.1 ' ' 1
```

(Dispersion parameter for quasipoisson family taken to be
    5.485093)

Number of Fisher Scoring iterations: 6

The syntax of the model in ❶ is straightforward. The object "mhp.survey"❷ contains the dataset and the information concerning the sampling design; it was constructed in Section 4.1. Results are presented in ❸; the interpretation is similar to that presented above. The exponential of the coefficients can be obtained from the instruction exp(coef(mod)) and the 95% confidence interval from exp(confint(mod)).

The negative binomial regression model can be obtained from the function glm.nb() in the package MASS:

❶

```
> mod <- glm.nb(nb.prison ~ char + ns + family.prison + abuse
    + age + f.type.centre, data = mhp.mod)
> summary(mod)
```

```
Call:
glm.nb(formula = nb.prison ~ char + ns + family.prison
    + abuse + age + f.type.centre, data = mhp.mod,
    init.theta = 2.32865466867588, link = log)
```

```
Deviance Residuals:
   Min        1Q      Median        3Q        Max
-1.8852   -0.7622    -0.4943    0.1736     4.5798

Coefficients:
                    Estimate    Std. Error    z value    Pr(>|z|)
(Intercept)        -0.143664      0.212235     -0.677    0.498464
char                0.221575❷     0.042560      5.206    1.93e-07 ***
ns                  0.142936      0.043809      3.263    0.001103 **
family.prison       0.379583      0.078049      4.863    1.15e-06 ***
abuse               0.288620      0.078345      3.684    0.000230 ***
age                 0.004558      0.002921      1.560    0.118739
f.type.centre2      0.159768      0.132489      1.206    0.227860
f.type.centre3      0.176061      0.129296      1.362    0.173296
---
Signif. codes:  0 '***' 0.001 '**' 0.01 '*' 0.05 '.' 0.1 ' ' 1

(Dispersion parameter for Negative Binomial(2.3287) family
   taken to be 1)

    Null deviance: 683.19 on 633 degrees of freedom
Residual deviance: 548.08 on 626 degrees of freedom
  (165 observations deleted due to missingness)
AIC: 2681.3

Number of Fisher Scoring iterations: 1

            Theta:   2.329
        Std. Err.:   0.211

2 x log-likelihood:   -2663.327
```

The syntax of the model is in ❶. The results (coefficients❷, standard deviations, p-values) are comparable to those obtained in the Quasi-Poisson regression model. It is, however, noteworthy that they are systematically lower in this example (by about 30%). The instructions exp(coef()), exp(confint()), drop1(), and plot(mod, 4) are also available in this context.

As explained in the introduction, the outcome variable "number of past imprisonments" is likely to present a "zero excess" phenomenon. To deal with this problem, a hurdle model can be used. It considers that the number of past imprisonments is the result of two processes: the first explains the binary variable "at least one past imprisonment" and the second process explains the number of past imprisonments when it is greater than 0 (Zeileis, Kleiber, and Jackman 2008). The function hurdle() in the package "pscl" can be used in this situation:

```
> mod <- hurdle(nb.prison ~ char + ns + family.prison + abuse
  + age + f.type.centre, data = mhp.mod, dist = "negbin")
> summary(mod)

Call:
hurdle(formula = nb.prison ~ char + ns + family.prison + abuse
  + age + f.type.centre, data = mhp.mod, dist = "negbin")
```

Count model coefficients (truncated negbin with log link):❶

	Estimate	Std. Error	z value	Pr(>|z|)	
(Intercept)	-2.156382	0.690875	-3.121	0.00180	**
char	0.328553	0.100562	3.267	0.00109	❷**
ns	0.172136	0.111964	1.537	0.12419	❸
family.prison	0.417507	0.180006	2.319	0.02037	*
abuse	0.466453	0.197573	2.361	0.01823	*
age	0.022593	0.009342	2.418	0.01559	*
f.type.centre2	0.762049	0.321729	2.369	0.01786	*
f.type.centre3	0.597374	0.305701	1.954	0.05069	.
Log(theta)	-0.970598	0.369799	-2.625	0.00867	**

Zero hurdle model coefficients (binomial with logit link):❹

	Estimate	Std. Error	z value	Pr(>|z|)	
(Intercept)	-0.898261	0.491763	-1.827	0.067758	.
char	0.159890	0.110106	1.452	0.146462	❺
ns	0.293772	0.099989	2.938	0.003303	❻**
family.prison	0.678375	0.190567	3.560	0.000371	***
abuse	0.568795	0.191324	2.973	0.002950	**
age	-0.001662	0.006782	-0.245	0.806342	
f.type.centre2	-0.423904	0.306814	-1.382	0.167084	
f.type.centre3	-0.121487	0.300311	-0.405	0.685819	

```
---
Signif. codes: 0 '***' 0.001 '**' 0.01 '*' 0.05 '.' 0.1 ' ' 1

Theta: count = 0.3789
Number of iterations in BFGS optimization: 20
Log-likelihood: -1085 on 17 Df
```

Results concerning the covariates associated with the number of past imprisonments (when there is at least one) are in ❶. Results concerning the outcome "at least one past imprisonment (yes/no)" are in ❹.

Concerning the variable "character" (the score reflecting the existence and the severity of an underlying personality disorder), it is associated with the number of past imprisonments❷ when there is at least one, but not with variable "at least one past imprisonment (yes/no)"❺. In other terms, a personality disorder does not appear to be associated with being a repeat offender or not, but is associated with being a multiple repeat offender. Exactly the reverse is true with the variable "novelty seeking" in ❸ and ❻.

6

About Statistical Modelling

> "Remember that all models are wrong; the practical question is how wrong do they have to be to not be useful" (Box and Draper 1987).

This well-known remark expresses the fact that statistical modelling is a compromise: a compromise between reliability and informativeness; a compromise between goodness of fit, adequacy of hypotheses on the one hand and, on the other, relevance and intelligibility of results (in particular, to the eyes of a non-statistician).

At the moment there are no consensual, clear-cut rules concerning the design of a good model—perhaps because sociologists, econometricians, and epidemiologists are likely to have different prerequisites about what is a good model. Perhaps because the compromise must be tailored to each particular study.

While no definite rules exist, some suggestions can nevertheless be made. In the previous chapter we saw that most models relate an outcome to a series of predictors. Depending on the nature of the outcome, a linear, logistic, or Poisson regression model will be chosen. In most regression models the outcome is associated with a linear combination of predictors, which makes computations and interpretations easier. But which predictors should be included? Should they be chosen *a priori*, as defined in a statistical analysis plan, or *a posteriori*, according to certain preliminary statistical analyses? Should the predictors be converted, by way of recoding or even use of products of predictors (also called interaction terms)? Alongside these questions, the problem of missing data also must be raised, as does the issue of the proper way to interpret the relative importance of a given predictor in a regression model. Finally, the notions of "resampling strategy" and "random effect" are now considered classic. What is the meaning of these sophisticated concepts?

These are issues that we are now going to consider in more detail.

6.1 Coding Numerical Predictors

In a few words: Let us consider the following linear regression model:

$$Y = a_0 + a_1 \times X_1 + a_2 \times X_2 + \ldots + a_p \times X_p + \varepsilon.$$

If the predictor X_i is the numerical variable "age", then an increase in X_i from 10 to 30 will have the same impact on Y as an increase from 60 and 80. The situation is, of course, comparable in a logistic regression or a Poisson regression.

Of course, most often, this hypothesis is not true. In some circumstances, it can even be completely meaningless: If Y is a cognitive functioning measurement, Y increases with age between 10 and 30 and decreases with age between 60 and 80.

One of the first stages in statistical modelling thus consists of the verification of the assumption of "linearity" between all numerical predictors and the outcome. This can be done using a regression spline. A spline, among other things, is a flexible strip of wood used to draw sweeping curves. In statistics, it corresponds to the best smooth curve that can summarize the relationship between two variables. If it is close to a straight line, the assumption of linearity is verified. If it is monotonic, with only a slight curvature, the assumption can be accepted but the statistical power of the test of association between Y and X_i will be depreciated. If the spline clearly has a non-linear shape (e.g., a U-shape), the model is not valid. In this situation, three solutions can be proposed:

1. To keep X_i as a spline in the model (but this no longer enables quantification of the strength of association between X_i and Y);
2. To convert X_i into a categorical variable with three, four, or more levels (but the model becomes less parsimonious and there are artificial skips when passing from one level to another); and
3. To add predictors like X_i^2, X_i^3, etc. (but results become more complex to interpret).

In Practice: In Chapter 5 we ran a linear regression model to explain the duration of interview according to several predictors, among which was the prisoner's age. But was there a linear relationship between these two variables? To answer this question we can draw a regression spline (see Figure 6.1) using the function gam() in the library "mgcv":

```
> library(mgcv)                    ❶
> mod <- gam(dur.interv ~ s(age), data = mhp.mod)
> plot(mod)
```

The only particularity in the syntax occurs in the function s() in ❶, which stands for "spline." The function gam() is useful because it is based on an algorithm that determines the level of complexity required to faithfully represent the relationship between Y and X. In Figure 6.1 ❷ it is specified that only one degree of freedom is required. The spline is thus linear (this can be easily seen in the figure). The predictor "age" can therefore be used confidently in the model.

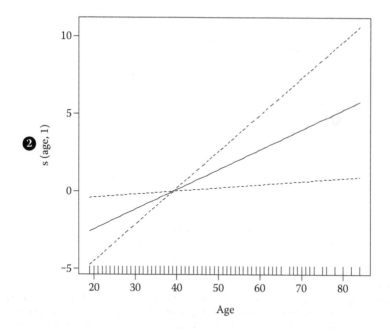

FIGURE 6.1
Regression spline representing the association between "duration of interview" and "age." The y-axis is coded in minutes (deviation from the mean duration of interview). The dashed lines are the 95% confidence bands of the spline. Because the spline is linear, the variable "age" can be used confidently in the linear regress.

In the case of logistic or Poisson regression modelling, splines can still be estimated using the function gam(). For example, if the outcome is now Y = "substance abuse and dependence (yes=1/no=0)" and the predictor is still "age," then the linearity between $Log\left[\dfrac{prob(Y=1)}{1-prob(Y=1)}\right]$ (which is termed logit[prob(Y=1)]) and "age" needs to be assessed. The syntax to use is

```
❶
> mod <- gam(subst.cons ~ s(age), data = mhp.mod,
             ❷
  family = binomial)
> plot(mod)
```

The syntax used is similar to the previous example. The instruction family = binomial❷ is used to specify that a logistic regression is required.
The spline is monotonic, slightly convex (see Figure 6.2). The estimated degree of freedom (i.e., the number of parameters required to obtain such a complex shape) is approximately equal to 3 in ❸. The y-axis gives the logit of the probability of a substance use disorder (in fact, it is the deviation from the mean logit).

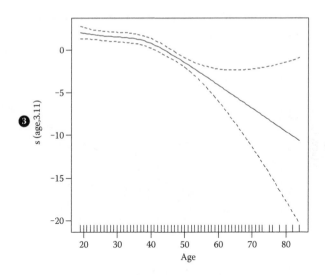

FIGURE 6.2
Regression spline representing the association between "substance abuse and dependence
(yes/no)" and age. The y-axis is coded as the logit of the probability of a substance use
disorder (deviation from the the mean logit). The spline is monotonic and slightly convex. The
hypothesis of linearity can be accepted but must be discussed.

The hypothesis of a linear relationship between "age" and logit[prob
(substance use disorder)] appears open to discussion. If one considers that
"age" is a predictor of major importance, it may be decided to use a more
sophisticated approach for its management in the logistic regression model.

The first possibility is to keep age as a spline in the model. For instance,
if "age" and "type of centre" are the two predictors of the model, then the
function gam() in the package "mgcv" gives the following results:

❶

```
> mod3 <- gam(subst.cons ~ s(age) + f.type.centre,
  data = mhp.mod, family = "binomial")
> summary(mod3)

Family: binomial
Link function: logit

Formula:
subst.cons ~ s(age) + f.type.centre

Parametric coefficients:
              Estimate Std. Error  z value Pr(>|z|)
(Intercept)    -2.6895     0.4074   -6.602 4.06e-11 ***
f.type.centre2  0.6324     0.3688    1.715 0.08637 .    ❷
f.type.centre3  1.0485     0.3467    3.024 0.00249 **

---
```

```
Signif. codes:  0 '***' 0.001 '**' 0.01 '*' 0.05 '.' 0.1 ' ' 1

Approximate significance of smooth terms: ❸
          edf     Ref.df    Chi.sq    p-value
s(age)    3.103   3.103     52.33     3.00e-11    ***❹
---
Signif. codes:  0 '***' 0.001 '**' 0.01 '*' 0.05 '.' 0.1 ' ' 1

R-sq.(adj) =   0.182 Deviance explained = 20.1%
UBRE score = -0.061167 Scale est. = 1 n = 797
```

The syntax concerning gam() is similar to that used for glm(), the main difference being the term s(age), which specifies that a spline is required for the variable "age." The results concerning the predictors are presented in ❷ and can be interpreted as in a glm() output. The results concerning the spline are in ❸; the p-value❹ indicates that "age" is significantly associated with a "substance abuse or dependence disorder." This approach is valuable from a theoretical point of view because it makes no assumptions concerning the shape of the relationship between "age" and the outcome. It, however, has a major drawback: There is no odds-ratio available for "age," and no way to quantify the strength of the relationship between the spline and the response.

Another possibility is to divide the variable "age" into, for instance, four pieces. One level is the reference, and three odds-ratios are estimated to assess to what extent each of the other three age categories differs in terms of probability of substance use disorder from the reference. The choice of the appropriate cut-offs is often challenging. They should appear "natural," with a well-balanced number of subjects in each category. In the present situation, 30, 40, and 50 years could be suggested. Unfortunately, because there is no prisoner over 50 years old with a substance abuse or dependence disorder, there is a risk of obtaining a non-finite odds-ratio for this category. The cut-offs 25, 35, and 45 years are therefore preferred:

❶
```
> mhp.mod$f.age <- cut(mhp.mod$age, breaks = c(-Inf, 25, 35,
    45, Inf), labels = c("young", "medyoung", "medold", "old"))
> table(mhp.mod$f.age, useNA = "ifany")❷

    young   medyoung   medold   old   <NA>
    140       217        202     238    2

> table(mhp.mod$subst.cons, mhp.mod$f.age, useNA = "ifany")❸

        young   medyoung   medold   old   <NA>
    0      71       133       150    232    1
    1      69        84        52     6     1
```

```
> mod3 <- glm(subst.cons~f.age + f.type.centre, data = mhp.mod,
  family = "binomial")
> summary(mod3)

Call:
glm(formula = subst.cons ~ f.age + f.type.centre,
  family = "binomial", data = mhp.mod)

Deviance Residuals:
   Min       1Q    Median       3Q      Max
-1.1875  -0.8570  -0.2626   1.1673   2.9809

Coefficients:
                  Estimate Std. Error z value Pr(>|z|)
(Intercept)        -1.0570     0.3886   -2.720 0.006525 **
f.agemedyoung      -0.2397❹    0.2274   -1.054 0.291920
f.agemedold        -0.8364❺    0.2405   -3.478 0.000505 ***
f.ageold           -3.3742❻    0.4537   -7.436 1.04e-13 ***
f.type.centre2      0.6805     0.3781    1.800 0.071901 .
f.type.centre3      1.0808     0.3554    3.041 0.002355 **
---
Signif. codes: 0 '***' 0.001 '**' 0.01 '*' 0.05 '.' 0.1 ' ' 1

(Dispersion parameter for binomial family taken to be 1)

    Null deviance: 921.27 on 796 degrees of freedom
Residual deviance: 757.98 on 791 degrees of freedom
  (2 observations deleted due to missingness)
AIC: 769.98

Number of Fisher Scoring iterations: 6

> exp(coefficients(mod3))
                (Intercept)    f.agemedyoung  f.agemedold    f.ageold
                 0.34748991       0.78687239   0.43327828  0.03424714❼
  f.type.centre2 f.type.centre3
     1.97482009     2.94699138
```

The subdivision of "age" is performed using the function cut() ❶. It is important to verify first that the number of prisoners is well balanced across the categories ❷, and that there are an adequate number of prisoners with a substance disorder in each of them ❸. The logistic regression is then performed on the categorical variable "f.age" (with, for the example, the addition of the covariate "type of centre"). It can be noted that the probability of a substance disorder decreases with age, in a steep and non-linear manner (❹, ❺, and ❻): A substance disorder is 30 times less frequent among the "old" prisoners than among the "young"❼ ($1/0.034 \approx 30$).

Another way of dealing with the non-linearity of age is to introduce squared and cubed terms in the model: logit[prob(substance use disorder)] $= a + b \times age + c \times age^2 + d \times age^3$. This can be easily implemented in the glm() function; however, there is a problem: age, age^2, and age^3 are strongly correlated, so that the interpretation of b, c, and d would be difficult. The simultaneous test of "$b = c = d = 0$" is possible; it will answer the following question: Is there a relationship between age and substance use disorder? However, as previously observed with spline coding, with this approach it is impossible to quantify the strength of association between the predictor and the outcome. A solution to this is to consider the orthogonal polynomials P_1 (degree 1), P_2 (degree 2), and P_3 (degree 3) instead of the single terms age, age^2, and age^3. These polynomials are designed so that P_1 efficiently captures the linear part of the association between age and the presence of a substance use disorder, P_2 captures the quadratic part of the association (i.e., the level of convexity present in a "U"-shape relationship), and P_3 captures a possible change in convexity, as in a "~"-shape relationship. In practice, the function poly() must be called in glm():

```
> mhp.int <-
  na.omit(mhp.mod[, c("subst.cons", "age", "f.type.centre")])
                              ❶
> mod4 <- glm(subst.cons ~ poly(age, degree = 3) +
  f.type.centre, data = mhp.int, family = "binomial")
> summary(mod4)

Call:
glm(formula = subst.cons ~ poly(age, degree = 3)
  + f.type.centre, family = "binomial", data = mhp.int)

Deviance Residuals:
  Min       1Q      Median     3Q       Max
-1.2474   -0.9151   -0.2973   1.1367   3.0075
```

Coefficients:

	Estimate	Std. Error	z value	Pr(>|z|)	
(Intercept)	-3.7647	0.7607	-4.949	7.46e-07	***
poly(age, degree = 3)1	-127.6394❷	39.7834	-3.208	0.00133	** ❸
poly(age, degree = 3)2	-92.3605❹	38.3662	-2.407	0.01607	* ❺
poly(age, degree = 3)3	-27.4751❻	16.4527	-1.670	0.09493	. ❼
f.type.centre2	0.6352	0.3817	1.664	0.09605	.
f.type.centre3	1.0482	0.3586	2.923	0.00347	**

```
---
Signif. codes:  0 '***' 0.001 '**' 0.01 '*' 0.05 '.' 0.1 ' ' 1

(Dispersion parameter for binomial family taken to be 1)
```

```
    Null deviance: 921.27 on 796 degrees of freedom
Residual deviance: 737.09 on 791 degrees of freedom
AIC: 749.1

Number of Fisher Scoring iterations: 9

Warning message:❽
In glm.fit(x = X, y = Y, weights = weights, start = start,
   etastart = etastart, : fitted probabilities numerically 0 or
   1 occurred
```

Because the degree of freedom of the spline was approximately equal to 3 in Figure 6.2, we now use orthogonal polynomials up to the third degree ❶. The linear trend is significant at the 5% level❸ (the older the prisoner, the smaller the likelihood of him presenting the disorder), as is the quadratic part ❺ (the probability is even smaller as prisoner age increases) but not the third-order polynomial ❼. Unfortunately, the corresponding coefficients ❷, ❹, and ❻ are difficult to interpret. The warning in ❽ corresponds to a situation of complete separation (Venables and Ripley 2002, p. 198). This is not a problem here, as it disappears when the non-significant third-degree orthogonal polynomial is discarded.

6.2 Coding Categorical Predictors

In a few words: In a questionnaire, most variables are of a categorical nature. Because regression models deal only with numbers, there is a need to recode numerically all categorical variables. Binary responses (yes/no, for instance) can be coded with "0" for "no" and "1" for "yes" and can be then considered predictors in a regression model. Ordered responses like "not satisfied," "satisfied," and "very satisfied" can be coded numerically in a somewhat natural way with, for example, "0" for "not satisfied," "1" for "satisfied," and "2" for "very satisfied." This numerical coding of course implies that the effect of the ordered variable on the outcome Y is similar when passing from "not satisfied" to "satisfied" to the effect when passing from "satisfied" to "very satisfied." If there is the smallest doubt concerning this assumption, it should be supported by analyses similar to those presented in the previous section.

When the answer to a question is neither binary nor ordered (like a profession, a brand of car, or a blood group), there is obviously a problem. Consider the ABO blood group system: Each human being is either A, B, AB, or O. If the blood group will be used as a predictor (named X_i) in the linear regression model $Y = a_0 + a_1 \times X_1 + \ldots + a_p \times X_p + \varepsilon$, how can this be done? If "A" is replaced by "1", "B" by "2", "AB" by "3", and "O" by "4", then it is explicitly assumed that the effect of B (here $a_i \times 2$) is exactly between the effect of A ($a_i \times 1$) and the effect of AB ($a_i \times 3$). Obviously this does not make sense, and no other coding can get rid of this paradoxical situation. A solution to this problem lies in the conversion of the variable "blood group" into three binary variables X_{i1}, X_{i2}, and X_{i3} (3 being the number of blood groups minus 1):

If blood group = A then $X_{i1} = 1$, $X_{i2} = 0$, $X_{i3} = 0$

If blood group = B then $X_{i1} = 0$, $X_{i2} = 1$, $X_{i3} = 0$

If blood group = AB then $X_{i1} = 0$, $X_{i2} = 0$, $X_{i3} = 1$

If blood group = O then $X_{i1} = 0$, $X_{i2} = 0$, $X_{i3} = 0$

These variables are then entered into the model:

$$Y = a_0 + a_1 \times X_1 + \ldots + a_{i1} \times X_{i1} + a_{i2} \times X_{i2} + a_{i3} \times X_{i3} + \ldots + a_p \times X_p + \varepsilon$$

From this new coding of "blood group", a_{i1} corresponds to the difference in Y between subjects with blood group O ($X_{i1} = 0$, $X_{i2} = 0$, $X_{i3} = 0$) and subjects with blood group A ($X_{i1} = 1$, $X_{i2} = 0$, $X_{i3} = 0$), given that the other predictor variables in the model are held constant. Similarly, a_{i2} corresponds to the difference in Y between subjects with blood group O and subjects with blood group B, and a_{i3} corresponds to the difference between subjects with blood group O and those with blood group AB. There is thus a common reference—here, blood group O.

Other codings are possible, even if the previous one (named "(0,1) coding") is the most popular. An alternative is the so-called "(1, –1) coding":

If blood group = A then $X_{i1} = 1$, $X_{i2} = 0$, $X_{i3} = 0$

If blood group = B then $X_{i1} = 0$, $X_{i2} = 1$, $X_{i3} = 0$

If blood group = AB then $X_{i1} = 0$, $X_{i2} = 0$, $X_{i3} = 1$

If blood group = O then $X_{i1} = -1$, $X_{i2} = -1$, $X_{i3} = -1$

In this situation, a_{i1} corresponds to the level of Y in subjects with blood group A, a_{i2} in subjects to blood group B, a_{i3} to blood group AB, and $-a_{i1} - a_{i2} - a_{i3}$ to blood group O. The reference is now the average of these four levels because $(a_{i1} + a_{i2} + a_{i3} - a_{i1} - a_{i2} - a_{i3})/4 = 0$

Fortunately, if an explanatory variable with p-levels is stated in R as a "factor," then the software automatically recodes it in the form of $(p - 1)$ binary variables in all the regression models that we have seen in the previous chapter.

Unfortunately, because this process is automatic, it is easy to forget the underlying recoding, and this can raise real problems in several circumstances. The following advice can be useful to avoid most of these problems:

- If an explanatory variable is a non-ordered, non-binary categorical variable, then it is a good habit to systematically check that it is entered as a "factor" (if not, the variable will be considered as numerical and the results will be uninterpretable).
- By default, R uses a (0,1) coding, which corresponds to the "contr. treatment" option in the R taxonomy.
- This (0,1) coding is satisfactory if the reference level makes sense. If not, this reference level should be changed.
- This (0,1) coding is satisfactory if the number of subjects in the level of reference is sufficient. Otherwise all comparisons will have a low statistical power and results will be difficult to interpret.
- This (0,1) coding should be used with extreme caution when certain interaction terms are included in the model (see Section 6.4).

In Practice: Let us consider the variable "profession" of the MHP study. This is typically a non-ordered, non-binary categorical variable. We first must verify how it is coded. The functions str() and summary() can be used for this purpose:

```
> str(mhp.mod$prof)
     ❶        ❷
  Factor w/ 8 levels "farmer", "craftsman", ..: 7 NA 4 6 8 6 7
  2 6 6 ...
> summary(mhp.mod$prof)
     farmer   craftsman    manager   intermediate   employee
        5❸         91         24           57          136
     worker      other      nojob         NA's
       228         31        221            6
```

The function str() explicitly mentions that "prof" is a "factor"❶ with eight levels❷. The function summary() details the factors and the number of prisoners in each profession. For instance, it can be seen that there are few farmers❸. The function describe() in the package "prettyR" could also have been used.

As in Section 5.2, let us imagine now that we are interested in explaining the binary outcome "at high risk for suicide attempt" by the following explanatory variables: "abuse" (yes/no), "disciplinary procedure" (yes/no), "duration of sentence" (<1 month, 1–6 months, 6–12 months, 1–5 years, >5 years), "age" (continuous), "prof" (categorical, see above), and "type of prison" (categorical 1, 2, 3). All these explanatory variables except "prof" and "type of prison" are considered numerical variables.

The function glm() gives the output:

```
> mod <- glm(suicide.hr ~ abuse + discip + duration + prof
  + f.type.centre, data = mhp.mod, family = "binomial")
> summary(mod)

Call:
glm(formula = suicide.hr ~ abuse + discip + duration + prof
  + f.type.centre, family = "binomial", data = mhp.mod)

Deviance Residuals:
   Min       1Q     Median      3Q        Max
-1.4691   -0.6910   -0.5482   -0.3343    2.4389

Coefficients:
                  Estimate   Std. Error   z value   Pr(>|z|)
(Intercept)        1.84840     1.50619      1.227    0.21975
abuse              0.62004     0.23521      2.636    0.00839  **
discip             0.39596     0.24881      1.591    0.11151
duration          -0.32932     0.14853     -2.217    0.02661  *
profcraftsman     -2.30497❶    1.28923     -1.788    0.07380  .❷
profmanager       -2.90743❸    1.44926     -2.006    0.04484  *❹
profintermediate  -3.05813     1.38046     -2.215    0.02674  *
profemployee      -2.54378     1.27738     -1.991    0.04644  *
profworker        -2.13259     1.25662     -1.697    0.08968  .
profother         -4.17093     1.62481     -2.567    0.01026  *
profnojob         -1.85382     1.25466     -1.478    0.13953
f.type.centre2    -0.06536     0.34254     -0.191    0.84868
f.type.centre3     0.31189     0.37935      0.822    0.41099
---
Signif. codes:  0 '***' 0.001 '**' 0.01 '*' 0.05 '.' 0.1 ' ' 1

(Dispersion parameter for binomial family taken to be 1)

    Null deviance: 554.57 on 546 degrees of freedom
Residual deviance: 512.92 on 534 degrees of freedom
  (252 observations deleted due to missingness)
AIC: 538.92
```

The variable "prof" has been automatically recoded in the form of seven binary variables (because there are eight professional categories). The coefficient a_{i1} relating to the first of these binary variables is in ❶, the corresponding p-value in ❷. This p-value corresponds to the test of the hypothesis: "given that the other predictor variables in the model are held constant, the profession 'craftsman' (i.e., artisan and self-employed) is not associated with a different probability of being at high risk for suicide attempt from that of the reference profession." The same applies for a_{i2} in ❸ and ❹. But what is the level of reference? It is the level omitted in the output, that is, "farmer." The output given by the function logistic.display() in the package "epicalc" provides results that are easier to read:

```
> library(epicalc)
> logistic.display(mod)
```

```
Logistic regression predicting suicide.hr❶
```

	crude	OR(95%CI)	adj.	OR(95%CI)	P(Wald's test)	P(LR-test)
abuse: 1 vs 0	1.87	(1.21,2.89)	1.86	(1.17,2.95)	0.008	0.009
discip: 1 vs 0	1.63	(1.03,2.57)	1.49	(0.91,2.42)	0.112	0.115
duration (cont. var.)	0.69	(0.55,0.87)	0.72	(0.54,0.96)	0.027	0.027
prof❷: ref. = farmer❸						0.012❻
craftsman	0.1	(0.01,1.19)	0.1❹	(0.01,1.25)	0.074	
manager	0.06	(0,0.92)	0.05❺	(0,0.94)	0.045	
intermediate	0.05	(0,0.68)	0.05	(0,0.7)	0.027	
employee	0.09	(0.01,1.1)	0.08	(0.01,0.96)	0.046	
worker	0.14	(0.01,1.57)	0.12	(0.01,1.39)	0.09	
other	0.03	(0,0.6)	0.02	(0,0.37)	0.01	
nojob	0.2	(0.02,2.27)	0.16	(0.01,1.83)	0.14	
f.type.centre: ref. = 1						0.376
2		1.06	(0.56,1.99)	0.94	(0.48,1.83)	0.849
3		1.79	(0.96,3.32)	1.37	(0.65,2.87)	0.411

```
Log-likelihood = -256.4597
No. of observations = 547
AIC value = 538.9193
```

The outcome is clearly presented in ❶. Concerning the explanatory variable "prof"❷, the level of reference is "farmer"❸. The adjusted odds-ratios (i.e., $\exp(a_{i1})$, $\exp(a_{i2})$, ...) are in ❹, ❺, etc. In ❻ we have the p-value of the global test that the variable "prof" as a whole has an influence on the probability of high risk for suicide attempt. This test is equivalent to the test that all the coefficients a_{i1}, a_{i2}, ... associated with the variable "prof" are simultaneously null. This test is obtained more classically using the function drop1() (the corresponding p-value is in ❼):

```
> drop1(mod, .~., test = "Chisq")
Single term deletions

Model:
suicide.hr ~ abuse + discip + duration + prof + f.type.centre
                  Df   Deviance   AIC      LRT    Pr(Chi)
<none>                 512.92   538.92
abuse             1    519.72   543.72   6.7974  0.009129 **
discip            1    515.40   539.40   2.4807  0.115250
duration          1    517.84   541.84   4.9206  0.026538 *
prof              7    530.83   542.83  17.9130  0.012369 *❼
f.type.centre     2    514.88   536.88   1.9568  0.375911
---
Signif. codes: 0 '***' 0.001 '**' 0.01 '*' 0.05 '.' 0.1 ' ' 1
```

At this stage, we can conclude that

1. The variable "prof" as a whole is significantly associated with the risk for suicide attempt;

2. The level of risk for "craftsman" is not different from the level of risk for "farmer"; while

3. The level of risk for "manager" appears to be significantly lower than the level of risk for "farmer" (and so forth for the other levels).

There are, however, some major concerns with these results. First, it is not obvious that "farmer" is the natural level of reference. Second, there are only six farmers in the MHP study. This implies that all the two-by-two comparisons presented above are made with a very small group of reference. The results are thus of limited interest, and the level of reference must be changed. The function relevel() can be used for this:

```
              ❶
> contrasts(mhp.mod$prof)
                ❷         ❸...
            craftsman manager intermediate employee worker other nojob
farmer❹         0       0          0           0       0      0     0
craftsman       1       0          0           0       0      0     0
manager         0       1          0           0       0      0     0
intermediate    0       0          1           0       0      0     0
employee        0       0          0           1       0      0     0
worker          0       0          0           0       1      0     0
other           0       0          0           0       0      1     0
nojob           0       0          0           0       0      0     1
```

```
                                            ❺
> mhp.mod$prof <- relevel(mhp.mod$prof, ref = "worker")
> contrasts(mhp.mod$prof)
             farmer craftsman manager intermediate employee other nojob
worker❻           0         0       0            0         0     0     0
farmer            1         0       0            0         0     0     0
craftsman         0         1       0            0         0     0     0
manager           0         0       1            0         0     0     0
intermediate      0         0       0            1         0     0     0
employee          0         0       0            0         1     0     0
other             0         0       0            0         0     1     0
nojob             0         0       0            0         0     0     1
```

The function contrasts()❶ can be used first to verify the coding of the "prof" variable. The binary variables $X_{i1}, X_{i2}, \ldots, X_{i7}$ used to recode "prof" are in ❷, ❸, etc. The level "farmer"❹ is the reference because all the a_{ij} ($j = 1$ to 7) are null. The function relevel() is used with the instruction ref = "worker"❺, which fixes the new level of reference. The function contrasts() can be used again to verify in ❻ that "worker" is indeed the new reference.

If glm() is called now with the same syntax, we obtain

```
> mod <- glm(suicide.hr ~ abuse + discip + duration + prof
  + f.type.centre, data = mhp.mod, family = "binomial")
> summary(mod)

Call:
glm(formula = suicide.hr ~ abuse + discip + duration + prof
  + f.type.centre, family = "binomial", data = mhp.mod)

Deviance Residuals:
   Min        1Q    Median       3Q      Max
-1.4691   -0.6910   -0.5482   -0.3343   2.4389


Coefficients:
                  Estimate Std. Error z value Pr(>|z|)
(Intercept)❶      -0.28420    0.78455  -0.362  0.71717
abuse❷             0.62004    0.23521   2.636  0.00839 **
discip             0.39596    0.24881   1.591  0.11151
duration          -0.32932    0.14853  -2.217  0.02661 *
proffarmer         2.13259    1.25662   1.697  0.08968 .❸
profcraftsman     -0.17238    0.40182  -0.429  0.66793   ❹
profmanager       -0.77484    0.77783  -0.996  0.31917
profintermediate  -0.92553    0.64236  -1.441  0.14963
profemployee❺     -0.41118    0.36301  -1.133  0.25734
profother         -2.03833    1.05821  -1.926  0.05408 .
profnojob❻         0.27877    0.26911   1.036  0.30024
```

```
f.type.centre2      -0.06536    0.34254    -0.191   0.84868
f.type.centre3❼     0.31189    0.37935     0.822   0.41099
---
Signif. codes: 0 '***' 0.001 '**' 0.01 '*' 0.05 '.' 0.1 ' ' 1

(Dispersion parameter for binomial family taken to be 1)

    Null deviance: 554.57 on 546 degrees of freedom
Residual deviance: 512.92 on 534 degrees of freedom
  (252 observations deleted due to missingness)
AIC: 538.92

Number of Fisher Scoring iterations: 5
```

In ❸, ❹, etc. are the p-values of the tests comparing each level to the new level of reference "worker." With this presentation, it is not possible to test, for example, the difference in risk for suicide attempt between the levels "no job" and "employee" (i.e., blue-collar worker). If another change in level of reference is a possibility ("no job" could be the new reference), there is another option using the function estimable() in the package "gmodels." This function can be used to test all linear combinations of the coefficients present in the previous output of glm(). There are 13 such coefficients, from the intercept❶, "abuse"❷, ..., up to the last "f.type.centre3"❼. We are interested in the 9th❺ and 11th❻ coefficients. We are, in fact, interested in a test that can compare ❺ – ❻ to zero. This can be done from the following syntax:

```
> library(gmodels)
                                           ❶         ❷
> estimable(mod, c(0, 0, 0, 0, 0, 0, 0, 0, 1, 0, -1, 0, 0))
                        Estimate   Std. Error X^2 value DF Pr(>|X^2|)
(0 0 0 0 0 0 0 1 0 -1 0 0) -0.6899548❸ 0.3607172 3.658539  1 0.05578253❹
```

The main point here is to determine carefully which comparison of coefficients should be tested. We are interested in the difference between the 9th and the 11th coefficient; so in estimable(), a vector of eight "0"s is specified (to discard the first 8 coefficients), one "1"❶ (to extract the coefficient of pro-femployee), one "0" (to discard the tenth coefficient), one "–1"❷ (to subtract the coefficient of profnojob), and two more "0"s (to discard the 12th and the 13th coefficients). The difference between the two coefficients is in ❸. The p-value of the test that compares this difference to 0 is in ❹; there is no difference between the level "no job" and the level "employee" in terms of risk for suicide attempt (given that the other predictor variables in the model are held constant).

From a numerical point of view, "worker" is a better reference level than "farmer" because there are only six farmers among the prisoners of the MHP study, while there are as many as 228 former unskilled workers. But from a scientific point of view, what would be the best profession of reference? There is no simple answer to this question. Perhaps a level of reference corresponding to the average effect of all eight types of profession would be more interesting here. This is what the (1,–1) coding obtained from the contr.sum option provides:

```
           ❶                              ❷
> contrasts(mhp.mod$prof) <- contr.sum
> contrasts(mhp.mod$prof)
              ❸ [, 1] [, 2] [, 3] [, 4] [, 5] [, 6] [, 7]
worker           1     0     0     0     0     0     0
farmer           0     1     0     0     0     0     0
craftsman        0     0     1     0     0     0     0
manager          0     0     0     1     0     0     0
intermediate     0     0     0     0     1     0     0
employee         0     0     0     0     0     1     0
other            0     0     0     0     0     0     1
nojob           -1    -1    -1    -1    -1    -1    -1
                      ❹
> colnames(contrasts(mhp.mod$prof)) <- c("worker", "farmer",
    "craftsman", "manager", "intermediate", "employee", "other")
```

The contrasts❶ for the variable "prof" are now estimated with the option "contr.sum"❷. There is no longer any level of reference (i.e., there is no line with seven "0"s). However, the sum (or equivalently, the mean) of the eight lines is equal to a line of "0"s. The average effect of the eight professions thus becomes the level of reference. Unfortunately, the seven binary variables X_{ij} (j = 1, …, 7) have lost their labels❸, and it may be useful to put them back as in ❹.

At this point, a new call for the glm() function gives

```
> mod <- glm(suicide.hr ~ abuse + discip + duration + prof
    + f.type.centre, data = mhp.mod, family = "binomial")
> summary(mod)

Call:
glm(formula = suicide.hr ~ abuse + discip + duration + prof
    + f.type.centre, family = "binomial", data = mhp.mod)

Deviance Residuals:
   Min       1Q     Median       3Q       Max
-1.4691   -0.6910   -0.5482   -0.3343    2.4389
```

```
Coefficients:
                   Estimate  Std. Error  z value  Pr(>|z|)
(Intercept)        -0.52306    0.82744    -0.632   0.52729
abuse               0.62004    0.23521     2.636   0.00839 **
discip              0.39596    0.24881     1.591   0.11151
duration           -0.32932    0.14853    -2.217   0.02661 *
profworker          0.23886❶   0.29659     0.805   0.42060 ❷
proffarmer          2.37146❸   1.10197     2.152   0.03140 *❹
profcraftsman       0.06649❺   0.39303     0.169   0.86567
profmanager        -0.53598❻   0.69640    -0.770   0.44151
profintermediate   -0.68667❼   0.58349    -1.177   0.23926
profemployee       -0.17232❽   0.36126    -0.477   0.63336
profother          -1.79947❾   0.93733    -1.920   0.05489 .
f.type.centre2     -0.06536    0.34254    -0.191   0.84868
f.type.centre3      0.31189    0.37935     0.822   0.41099
---
Signif. codes:  0 '***' 0.001 '**' 0.01 '*' 0.05 '.' 0.1 ' ' 1

(Dispersion parameter for binomial family taken to be 1)

    Null deviance: 554.57 on 546 degrees of freedom
Residual deviance: 512.92 on 534 degrees of freedom
    (252 observations deleted due to missingness)
AIC: 538.92

Number of Fisher Scoring iterations: 5
```

The coefficient ❶ is positive, showing that workers are numerically at a higher risk for suicide attempt than the average of all professions. However, this increase in risk is not statistically significant because the corresponding p-value ❷ is greater than 0.05. Farmers, however, are at even higher risk❸, and this is statistically significant in ❹.

The (1,–1) coding can be appealing in some circumstances, very notably when it makes sense to compare each level to the group as a whole. It has at least two drawbacks, however. First, it is not used frequently and can thus be misleading for a non-specialist readership. Second, in the glm() output, one level is lacking: "no job" in our example. If one is interested solely in the magnitude of the coefficient, it can easily be obtained from the other one because their summation is equal to 0 (the missing coefficient is thus equal to – ❶ – ❸ – ❺ – ❻ – ❼ – ❽ – ❾). If one is also interested in the p-value of the test that compares this missing level to the average of all eight professions, then the function estimable() in the package "gmodels" must be used again:

```
> library(gmodels)
> estimable(mod, c(0, 0, 0, 0, -1, -1, -1, -1, -1, -1, -1, 0, 0),
  conf.int = 0.95)
                                 Estimate    Std. Error X^2 value DF
(0 0 0 0 -1 -1 -1 -1 -1 -1 -1 0 0)  0.5176350❶  0.2952967   3.072771  1
                                 Pr(>|X^2|)  Lower.CI    Upper.CI
(0 0 0 0 -1 -1 -1 -1 -1 -1 -1 0 0)  0.07961365❷ -0.06822514 1.103495
```

The missing coefficient is in ❶, and the p-value in ❷; the level "no job" is at
higher risk than the average of all eight jobs (❶ is positive) but this increase
is not significant at the 5% level (❷ is greater than 5%).

6.3 Choosing Predictors

In a few words: Which explanatory variables should be included in a model?
This day-to-day question has been the subject of much discussion (Harrel
2001) and little consensus, perhaps because all studies cannot be approached
in the same way.

Traditionally, the job of a scientist is supposed to be to formulate a hypothesis and to perform a reproducible experiment, the aim of which is to refute
or support that hypothesis. Basically, science is thus presented as a confirmatory process. This view is, however, not totally in accordance with what a
scientist does in his (her) real life, especially when his (her) job is to analyse
questionnaires. There are indeed situations where there is a single hypothesis
and a corresponding analysis to support it, but these situations are not very
frequent. In contrast, there are also totally exploratory situations, where the
researcher has no or few ideas about what he (she) is looking for. Most often,
however, the situation is intermediate: The scientist has a series of hypotheses
defined *a priori* (but not always explicitly and clearly formulated) and he (she)
is open to results that may not necessarily be anticipated. In each particular
case, the question of choosing appropriate predictors in a regression model
will receive a different answer. Let us now see a few typical situations.

If it is rare for a questionnaire survey to be designed to answer a single
question, it can happen that a very precise scientific problem is raised and
that it finds a solution in the analysis of data collected through a vast questionnaire survey. In this situation, there is in general, on the one hand, an
outcome (the dependant variable Y) and, on the other hand, a main predictor
of interest (X_1). If a basic t-test or chi-square test is essential here (to test the
basic association between Y and X_1), then a regression model can still be of
great interest to deal with other predictors that are likely to raise problems
in the interpretation of the relationship between Y and X_1. These secondary predictors should be decided *a priori*, on scientific grounds (e.g., after a

review of the literature discussing the confounding factors that could affect the relationship between Y and X_1). Because this choice can be difficult in practice, one solution sometimes proposed (Dalton et al. 2003) is to focus on two models instead of one: a first model with the two, three, and or four essential secondary predictors and a second model that includes most of them. The first model will be easier to read but will bring to light a rather rough version of the relationship between Y and X_1. The second model will be, in general, more obscure but also more specific concerning the pair of variables of interest. Of course, the condition of the validity of each model must be met; this often implies a limitation on the number of predictors that can be put into the model (see Section 5.2 for details).

This approach can be generalized when there is a limited list of predictors of interest $X_1, \ldots, X_q$ instead of a unique predictor of interest X_7.

Sometimes the situation is less clear. There is no longer a single predictor or a limited series of predictors, but instead a rather long and partly redundant list of predictors and the first step is to reduce it. Variables with many missing data, or with a small variance (i.e., questions that often received the same answer and that are thus poorly informative) are ideal candidates for removal. Redundant variables can also be removed, and they can be checked using exploratory multidimensional data analysis (see Sections 3.5, 3.6, and 3.7). Prior knowledge about the relevance of the relationship between the outcome Y and a given predictor can also be used to discard the latter.

At this stage, if the list of predictors of interest is short enough (i.e., compatible with the sample size, the type of model used, and its readability), then the model can be estimated. If not, other options must be considered, but there will be a price to pay in most cases. that is, a certain lack of trustworthiness in the results.

Among these other options, a classic one is to look at the basic bivariate relationship between the outcome Y and each of the potential predictors $X_1, \ldots, X_p$ (using t-test, chi-square test, or correlation tests) and to select the predictors with the smallest p-values. Another classic option is to let the software automatically construct the model, entering and withdrawing step by step the most and least significant predictors. This approach is often called a "stepwise model selection" procedure. Both of these options (especially the second) have a major drawback: they lead to over-optimistic models. Imagine there is a list of 200 potential predictors that are, in fact, not associated with the outcome Y: about 5% of them (i.e., 10) are likely to be nevertheless associated "by chance" to Y in the sample under study. If a regression model relates Y to these 10 predictors, this model will appear deceptively interesting (especially if it is not mentioned that the 10 predictors where selected from a list of 200). From a formal point of view, coefficients and p-values of the predictors found in the final stage of a stepwise selection procedure are biased and this bias can be considerable (Harrel 2001). Stepwise selection procedures should therefore be used only when no other approaches are applicable, and they should be mainly considered as exploratory analyses.

In Practice: Imagine that, in the MHP study, we are interested in a regression model where the outcome is "being at high risk for suicide attempt" (Y) and the two main predictors are "childhood abuse" (yes[1]/no[0]) and "disciplinary procedure" (yes[1]/no[0]).

As explained above, a first approach to designing this regression model is, in fact, to consider two models: one minimally adjusted and one fully adjusted model. It is an *a priori* decision that the short list of secondary predictors for the minimal model will be "age" (numerical) and "type of centre" (categorical coded in "1", "2", and "3"), and the extensive list for the full model: "age" (numerical), "type of centre" (categorical coded in "1", "2", "3"), "separation from parents during childhood" (yes[1]/no[0]), "number of children" (numerical), "number of siblings" (numerical), "profession before prison" (yes[1]/no[0]), "remand prisoner" (yes[1]/no[0]), and the three temperament variables "harm avoidance," "reward dependence," and "novelty seeking" (numerical). All these variables could explain, at least in part, an association between "being at risk for suicide attempt" and "childhood abuse" or "disciplinary procedure." We obtain finally the following logistic regression models:

❶

```
> mhp.int <- na.omit(mhp.mod[, c("suicide.hr", "n.child",
  "n.brother", "prof.yn", "remand", "discip", "separation",
  "abuse", "age", "ns", "ha", "rd", "f.type.centre")])
> dim(mhp.int)
```
❷
```
[1] 609 13
> mod1 <- glm(suicide.hr ~ abuse + discip + age + f.type.centre,
  data = mhp.mod, family = "binomial")
> summary(mod1)

Call:
glm(formula = suicide.hr ~ abuse + discip + age + f.type.centre,
  family = "binomial", data = mhp.mod)

Deviance Residuals:
  Min       1Q      Median      3Q       Max
-1.1044   -0.6903   -0.5952   -0.4565   2.2276


Coefficients:
                  Estimate  Std. Error  z value  Pr(>|z|)
(Intercept)      -1.423968   0.468889    -3.037   0.00239  **
abuse             0.775780   0.193334     4.013   6e-05    ***❸
discip            0.320294   0.217106     1.475   0.14013       ❹
age              -0.014207   0.008145    -1.744   0.08112  .
f.type.centre2    0.095687   0.327819     0.292   0.77037
f.type.centre3    0.423626   0.311543     1.360   0.17390

---
```

```
Signif. codes: 0 '***' 0.001 '**' 0.01 '*' 0.05 '.' 0.1 ' ' 1

(Dispersion parameter for binomial family taken to be 1)

    Null deviance: 758.85 on 749 degrees of freedom
Residual deviance: 729.14 on 744 degrees of freedom
  (49 observations deleted due to missingness)
AIC: 741.14

Number of Fisher Scoring iterations: 4
```

A first important point is to be sure that the same dataset will be used for the different analyses. This is not trivial, as there are missing data and not all models will involve the same variables. The function na.omit()❶ is therefore used to select prisoners having no missing data for the fully adjusted model.

The second point involves the assessment of the sample size available for these analyses. There are 609❷ prisoners with no missing data.

The logistic regression indicates then that "childhood abuse" is significantly associated with "being at high risk for suicide attempt"❸, while this is not the case for "disciplinary procedure"❹.

In this model, however, we had a small number of adjustment variables. The fully adjusted model gives the following results:

```
> mod2 <- glm(suicide.hr ~ abuse + separation + discip + age
  + n.child + n.brother + school + prof.yn + remand + ns + ha
  + rd + f.type.centre, data = mhp.mod, family = "binomial")
> summary(mod2)

Call:
glm(formula = suicide.hr ~ abuse + separation + discip + age
  + n.child + n.brother + school + prof.yn + remand + ns + ha
  + rd + f.type.centre, family = "binomial", data = mhp.mod)

Deviance Residuals:
   Min       1Q     Median       3Q        Max
-1.7305   -0.6672   -0.5053   -0.3492    2.3853

Coefficients:
              Estimate  Std. Error  z value  Pr(>|z|)
(Intercept)  -3.472699    0.869573   -3.994  6.51e-05 ***
abuse         0.506381    0.239746    2.112    0.0347 *❺
separation    0.007652    0.229617    0.033    0.9734
discip        0.292272    0.254265    1.149    0.2504   ❻
age          -0.010032    0.010816   -0.928    0.3536
n.child       0.033430    0.068127    0.491    0.6236
```

```
n.brother         0.073085   0.031357    2.331   0.0198  *
school           -0.062049   0.125617   -0.494   0.6213
prof.yn          -0.353056   0.233431   -1.512   0.1304
remand            0.219512   0.250198    0.877   0.3803
ns                0.330152   0.129256    2.554   0.0106  *
ha                0.524955   0.129700    4.047   5.18e-05 ***
rd                0.023268   0.137648    0.169   0.8658
f.type.centre2   -0.131726   0.383583   -0.343   0.7313
f.type.centre3    0.009549   0.383055    0.025   0.9801
---
Signif. codes:  0 '***' 0.001 '**' 0.01 '*' 0.05 '.' 0.1 ' ' 1

(Dispersion parameter for binomial family taken to be 1)

    Null deviance: 598.37  on 607  degrees of freedom
Residual deviance: 543.31  on 593  degrees of freedom
    (191 observations deleted due to missingness)
AIC: 573.31

Number of Fisher Scoring iterations: 4
```

The same tendency is observed in ❺ and ❻ as in ❸ and ❹; this supports the previous conclusion obtained in the minimally adjusted model. Of course, the condition of the validity of these two models should be verified, as explained in Section 5.2.

Now, if one has definitely no *a priori* idea concerning the predictors of interest, a stepwise regression model could be attempted. The function step() can be used for this:

```
                            ❶
> mhp.int <- na.omit(mhp.mod[, c("suicide.hr", "grav.cons",
   "bipdis.cons", "dep.cons", "man.cons", "pan.cons",
   "ago.cons", "socp.cons", "ocd.cons", "ptsd.cons",
   "gad.cons", "alc.cons", "subst.cons", "psychosis.cons",
   "scz.cons", "sczdys.cons", "deldis.cons", "n.child",
   "n.brother", "school", "prof.yn", "remand", "against.goods",
   "against.people", "work.prison", "discip", "past.prison",
   "child.judge", "out.placed", "death", "separation", "abuse",
   "trauma", "family.prison", "age", "direct", "coop",
   "emostab", "transc", "ns", "ha", "rd", "suicide.past",
   "f.type.centre")])
> dim(mhp.int)
     ❷
[1] 510 44
        ❸
> mod1 <- glm(suicide.hr ~ abuse + discip + age + f.type.centre,
   data = mhp.int, family = "binomial")
```

❹
```
> mod.stp <- step(mod1, scope = list(upper = ~grav.cons
  + bipdis.cons + dep.cons + man.cons + pan.cons + ago.cons
  + socp.cons + ocd.cons + ptsd.cons + gad.cons + alc.cons
  + subst.cons + psychosis.cons + scz.cons + sczdys.cons
  + deldis.cons + n.child + n.brother + school + prof.yn
  + remand + against.goods + against.people + work.prison
  + discip + past.prison + child.judge + out.placed + death
  + separation + abuse + trauma + family.prison + age + direct
  + coop + emostab + transc + ns + ha + rd + suicide.past
  + f.type.centre, lower = ~f.type.centre❺), trace = FALSE)
> summary(mod.stp)

Call:
glm(formula = suicide.hr ~ f.type.centre + grav.cons
  + suicide.past + dep.cons + direct + past.prison
  + sczdys.cons + death + against.people + subst.cons,
  family = "binomial", data = mhp.int)

Deviance Residuals:
   Min        1Q      Median       3Q        Max
-2.03270   -0.41610   -0.18017   -0.06366   3.00219

Coefficients:
                 Estimate  Std. Error  z value  Pr(>|z|)
(Intercept)      -4.07548    1.10077    -3.702   0.000214 ***
f.type.centre2   -0.45153    0.50735    -0.890   0.373479
f.type.centre3   -0.08397    0.46080    -0.182   0.855399
grav.cons         0.68482    0.14987     4.569   4.89e-06 ***
suicide.past      2.02125❻   0.33153     6.097   1.08e-09 ***❼
dep.cons          1.35671    0.32983     4.113   3.90e-05 ***
direct            1.17913❽   0.34609    -3.407   0.000657 ***
past.prison       1.02016    0.34186     2.984   0.002844 **
sczdys.cons       1.43947    0.82453     1.746   0.080845 .
death             0.59614    0.33465     1.781   0.074853 .
against.people   -0.60321    0.35030    -1.722   0.085073 .
subst.cons       -0.55316    0.36347    -1.522   0.128042
---
Signif. codes: 0 '***' 0.001 '**' 0.01 '*' 0.05 '.' 0.1 ' ' 1

(Dispersion parameter for binomial family taken to be 1)

    Null deviance: 507.63 on 509 degrees of freedom
Residual deviance: 273.55 on 498 degrees of freedom
AIC: 297.55

Number of Fisher Scoring iterations: 6
```

Once again, prisoners with missing data are discarded❶ to obtain a single dataset for the forthcoming analyses. There are only 510 prisoners remaining❷. A basic model is estimated in ❸; it will be used as a starter in the step() procedure. There is an "upper" list of potential predictors❹ (all the variables that can be entered into the model) and a "lower" list❺ (the variables that will be forced into the model). The final model mod.stp is different from the two models estimated previously. The variable "past episode of suicide attempt" is the strongest❻ and most significant❼ predictor associated with the outcome.

Which model is best: this last one, the minimally adjusted model, or the fully adjusted one? In fact, they follow different logics.

The minimally and fully adjusted models should be considered together. They are based on the (fallible) prior hypotheses of the investigator and obey a fairly strict hypothetical and deductive procedure. However, because they are designed *a priori*, these models can miss important aspects of the question. For this reason, it is conceivable to modify them punctually during the analysis if a limited list of unexpected predictors appears to be of major interest. This is not to say that all statistically significant predictors must be included in a model. For example, we can consider that "childhood abuse" explains "past episode of suicide attempt," which in turn explains the outcome "being at risk of suicide attempt." If "childhood abuse" and "past episode of suicide attempt" are both in the model, then "childhood abuse" will be masked and will appear unrelated to Y. Even if this is correct from a statistical point of view, it can be misleading if results are examined too cursorily. If one is particularly interested in the relationship between childhood abuse and risk of suicide attempt, then "past episode of suicide attempt" should be discarded.

The stepwise regression presents a series of predictors that are strongly associated with the outcome. However, the coefficients and p-values are biased and there is no theoretical perspective behind the selection process. Results are thus mainly exploratory.

There is an efficient but costly way of dealing with the drawbacks of stepwise model selection: to divide the sample into two parts, the first being used to develop the model, the second to estimate the coefficients and the p-values. This can be done easily using the function sample():

```
                    ❶                    ❷
> dt.trial <- sample(1:511, size = 255)
                    ❸
> mhp.trial <- mhp.int[dt.trial, ]
                    ❹
> mhp.conf <- mhp.int[-dt.trial, ]
> mod <- glm(suicde.hr ~ abuse + discip + age + f.type.centre,
          ❺
     data = mhp.trial, family = "binomial")
```

```
> mod.stp <- step(mod, scope = list(upper = ~grav.cons
  + bipdis.cons + dep.cons + man.cons + pan.cons + ago.cons
  + socp.cons + ocd.cons + ptsd.cons + gad.cons + ....
  [TRUNCATED]
> summary(mod.stp)

Call:
glm(formula = suicide.hr ~ discip + f.type.centre + grav.cons
  + suicide.past + direct + dep.cons + past.prison + bipdis.cons
  + school + gad.cons + remand, family = "binomial",
  data = mhp.trial)

Deviance Residuals:
Min         1Q          Median      3Q          Max
-1.78439    -0.27730    -0.05939    -0.01013    2.83622

Coefficients:
                Estimate  Std. Error  z value  Pr(>|z|)
(Intercept)     -9.40730  2.52079     -3.732   0.00019   ***
discip          -1.17515  0.67889     -1.731   0.08345   .❻
f.type.centre2  -0.90213  0.82278     -1.096   0.27288
f.type.centre3   0.04453  0.76871      0.058   0.95380
grav.cons        1.26019  0.30651      4.111   3.93e-05  ***
suicide.past     2.83206  0.63456      4.463   8.08e-06  ***
direct           2.07190  0.63577     -3.259   0.00112   **
dep.cons         1.42398  0.54673      2.605   0.00920   **
past.prison      1.66490  0.59936      2.778   0.00547   **
bipdis.cons     -2.17835  1.13436     -1.920   0.05482   .
school           0.59426  0.28594      2.078   0.03769   *
gad.cons         1.07447  0.55278      1.944   0.05193   .
remand           0.96354  0.59072      1.631   0.10286
---
Signif. codes: 0 '***' 0.001 '**' 0.01 '*' 0.05 '.' 0.1 ' ' 1

(Dispersion parameter for binomial family taken to be 1)

    Null deviance: 246.66 on 254 degrees of freedom
Residual deviance: 109.83 on 242 degrees of freedom
AIC: 135.83

Number of Fisher Scoring iterations: 7

                    ❼                     ❽
> mod <- glm(mod.stp$formula, data = mhp.conf,
  family = "binomial")
> summary(mod)
```

```
Call:
glm(formula = mod.stp$formula, family = "binomial",
    data = mhp.conf)

Deviance Residuals:
    Min          1Q      Median          3Q         Max
-1.98199    -0.39638    -0.16701    -0.06534     2.96540

Coefficients:
                   Estimate Std. Error  z value  Pr(>|z|)
(Intercept)        -4.5638     1.8478    -2.470   0.01352  *
discip              1.1321     0.5281     2.144   0.03206  *❾
f.type.centre2     -0.1514     0.7826    -0.193   0.84664
f.type.centre3      0.2356     0.7212     0.327   0.74394
grav.cons           0.6292     0.1982     3.175   0.00150  **
suicide.past        2.6659     0.5132     5.194   2.05e-07 ***
direct              0.4193     0.5242    -0.800   0.42374
dep.cons            1.2819     0.4723     2.714   0.00665  **
past.prison         0.6793     0.4608     1.474   0.14042
bipdis.cons         0.6326     0.7167     0.883   0.37742
school             -0.4790     0.2988    -1.603   0.10883
gad.cons           -0.7938     0.5196    -1.528   0.12660
remand             -0.1329     0.5298    -0.251   0.80200
---
Signif. codes:  0 '***' 0.001 '**' 0.01 '*' 0.05 '.' 0.1 ' ' 1
```

Among numbers running from 1 to 511❶ (511 is the number of prisoners without missing data), 255 are selected randomly❷. The corresponding prisoners constitute the "trial" sample❸, the others constitute the "confirmatory" sample❹. The stepwise model selection is performed on the trial sample❺. The variables retained in the final step are "discip" (disciplinary procedure, yes/no), "f.type.centre" (type of prison), "grav.cons" (consensus for gravity, from 1 to 7), "suicide.past" (past history of suicide attempt), "direct" (low self-directedness, yes/no), "dep.cons" (depression, yes/no), "past.prison" (past history of imprisonment, yes/no), "bipdis.cons" (bipolar disorder, yes/no), "school" (school level, from 1 to 5), "gad.cons" (generalized anxiety disorder, yes/no), and "remand" (remand prisoner, yes/no).

This model is then estimated in the second independent sample❽. Note the instruction mod.stp$formula❼ that is used to call the model obtained at the end of the stepwise selection.

A cross-validation of this sort is not a good idea here because the two samples are too small. It is nevertheless interesting to take a look at the remarkable differences between the two series of results, despite the fact that they are derived from the estimation of the same model. For instance, "disciplinary procedure" is negatively associated with the outcome and p = 0.083❻ in the trial sample, while it is positively associated to the outcome with p = 0.032❾ in the confirmatory sample.

6.4 Interaction Terms

In a few words: In the different types of regression models we saw in Chapter 5, the effects of the predictors X_i are added together to explain the outcome Y. This is also the case in the following logistic regression model where Y is "laryngeal cancer" (yes[1]/no[0]), X_1 is "alcohol consumption" (yes[1]/no[0]), and X_2 is "tobacco consumption" (yes[1]/no[0]): $Log\left[\dfrac{prob(Y=1)}{1-prob(Y=1)}\right]$ $= a_0 + a_1 \times X_1 + a_2 \times X_2$. This relationship has a non-trivial implication: The effect of alcohol on the risk of presenting a laryngeal cancer is the same whether or not a subject smokes. Unfortunately, this does not correspond to what is commonly observed (Flanders and Rothman 1982): "The exposure to both factors increases the risk about 50% more than the increase predicted if the effects of tobacco and alcohol were simply additive." In such a situation, it is said that there is an "interaction" between alcohol and tobacco and that this interaction is positive because the exposure to both factors increases the risk.

From a statistical point of view, a simple way to capture certain elements of interaction between X_1 and X_2 consists of adding the product $X_1 \times X_2$ to the model. In our previous example, this now gives the following:

$$Log\left[\frac{prob(Y=1)}{1-prob(Y=1)}\right] = a_0 + a_1 \times X_1 + a_2 \times X_2 + a_3 \times X_1 \times X_2,$$

so that in smokers ($X_2 = 1$), we have

$$Log\left[\frac{prob(Y=1)}{1-prob(Y=1)}\right] = a_0 + a_2 \times 1 + a_1 \times X_1 + a_3 \times X_1 \times 1 = (a_0 + a_2) + (a_1 + a_3) \times X_1$$

and in non-smokers ($X_2 = 0$), we have

$$Log\left[\frac{prob(Y=1)}{1-prob(Y=1)}\right] = a_0 + a_1 \times X_1.$$

There is now a different level of association between the risk of presenting laryngeal cancer and alcohol consumption, depending on whether or not a subject smokes. In practice, following this approach, the test of the hypothesis $a_3 = 0$ will formally answer the following question: Is there a statistically significant interaction between X_1 and X_2?

Two problems are commonly encountered in the management of interaction terms in statistical models. The first concerns the identification of pairs of variables for which an interaction should be considered. It is often a disconcerting and unmanageable task to look carefully for each possible case. Here again, a review of the literature on the particular topic addressed by the model will help find the main potential interactions. An interesting approach proposed by Cox and Wermuth (1996) is presented in the practical part of this section. It helps determine globally whether or not a set of outcomes and predictors raises the question of interaction terms.

The second problem concerns the interpretation of the so-called main effects X_1 (alcohol consumption) and X_2 (tobacco consumption) in the presence of the interaction $X_1 \times X_2$. This point is regularly discussed in the literature (Fox 2002; Rothman and Greenland 1998) and there is no real consensus. Here are some elements of this debate:

- In our example, the test of $a_1 = 0$ involves the test of the effect of alcohol consumption on the risk of presenting a laryngeal cancer in a population of non-smokers (because in this situation $X_2 = 0$ so that the interaction term disappears). This interpretation is possible because $X_2 = 0$ has a clear meaning here. This is not always the case, for example if the variable X_1 was the "age": age = 0 is birth, which is not necessarily a relevant reference. In this latter situation, it is sometimes suggested (Rothman and Greenland 1998) that "age" should be re-centreed, for example by using the relation "age2 = age − 18" or "age2 = age − 65," if the age of 18 or 65 is considered a milestone of the disease (depending on the question under study).

- The previous option has a drawback: The test of $a_1 = 0$ does not involve the effect of alcohol consumption "in general" but only in a particular population. This is correct from a statistical point of view, but it can be frustrating from a public health perspective. An alternative sometimes proposed is to centre X_1 and X_2 by their mean; more precisely, X_1 is replaced by $X'_1 = X_1 - \text{mean}(X_1)$ and X_2 is replaced by $X'_2 = X_2 - \text{mean}(X_2)$. In this situation, the test of $a_1 = 0$ concerns the effect of alcohol consumption in a virtual population of subjects with a prevalence of tobacco consumption similar to the prevalence observed in the study (Koopman 2008). The interpretation is perhaps more intuitive, but this approach implies a manipulation of data and it is difficult to generalize to non-binary categorical variables.

- A third possibility (which is applicable to categorical variables) is to recode variables X_1 and X_2 as (1,−1) instead of (1,0) or, more generally and following the R taxonomy, to use a "contr.sum" option instead of the traditional "contr.treatment" one. Here, the test of $a_1 = 0$ involves the effect of alcohol consumption in a virtual population of subjects with a prevalence of tobacco consumption of 50%.

- Finally, some authors are more straightforward and suggest estimating the main effects using a model where the interaction term has been simply discarded (Fox 2002). This makes sense, even if it raises a theoretical problem: If the interaction does exist and if the interaction term is not in the model, then it is in the residual. But in this case, the residual is no longer "pure noise" (if the residual contains the interaction term $X_1 \times X_2$, then it is not independent of X_1 and X_2).

In Practice: In the previous section we designed a series of models that explained the probability of being at high risk for suicide attempt. If, for instance, we consider the model obtained using the stepwise selection procedure, 10 explanatory variables were selected. Among them was the variable named "direct," which stands for a low self-directedness (yes[1]/no[0]; low self-directedness is defined by poor impulse control or a weak ego (Cloninger 2000)). The variable "direct" was positively and significantly associated with the outcome, the odds-ratio being equal to exp(1.18❸) = 3.3). Another variable was named "f.type.centre," which stands for the type of prison: high-security unit (type 1), centre for longer sentences or for prisoners with good scope for rehabilitation (type 2), or a prison for remand prisoners or short sentences (type 3). This variable was not significantly associated with the outcome. It could be hypothesised that low self-directedness is particularly deleterious in high-security prisons as regards risk of suicide attempt, so that this could correspond formally to an interaction between "direct" and "f.type. centre" in explaining "suicide.hr."

Let us now see how to estimate and test this interaction in the corresponding logistic regression model:

```
                          ❶
> str(mhp.mod$direct)
 num [1:799] 0 0 0 0 0 0 0 0 1 0 ...
                          ❷
> contrasts(mhp.mod$f.type.centre)
      2   3
1     0   0❸
2     1   0
3     0   1
> mod <- glm(suicide.hr ~ grav.cons + suicide.past + dep.cons
  + past.prison + sczdys.cons + death + against.people
  + subst.cons + direct + f.type.centre
                  ❹
  + direct:f.type.centre, data = mhp.mod, family = "binomial")
> summary(mod)

Call:
glm(formula = suicide.hr ~ grav.cons + suicide.past + dep.cons
   + past.prison + sczdys.cons + death + against.people
   + subst.cons + direct + f.type.centre + direct:f.type.centre,
   family = "binomial", data = mhp.mod)
```

```
Deviance Residuals:
Min       1Q         Median    3Q        Max
-2.03462  -0.43702   -0.19137  -0.07424  3.16051
```

```
Coefficients:
                    Estimate Std. Error z value Pr(>|z|)
(Intercept)         -6.0626     0.7330   -8.271  < 2e-16 ***
grav.cons            0.6692     0.1207    5.545  2.94e-08 ***
suicide.past         1.7729     0.2791    6.352  2.13e-10 ***
dep.cons             1.4319     0.2893    4.950  7.43e-07 ***
past.prison          0.7321     0.2867    2.554  0.0107 *
sczdys.cons          1.3722     0.8377    1.638  0.1014
death                0.3008     0.2868    1.049  0.2943
against.people      -0.2359     0.2984   -0.791  0.4292
subst.cons          -0.2405     0.3100   -0.776  0.4380
direct              -1.4466❺    1.0311   -1.403  0.1606
f.type.centre2      -0.5643     0.4960   -1.138  0.2553
f.type.centre3      -0.3066     0.4624   -0.663  0.5073
direct:f.type.centre2  2.8487❻  1.1701    2.435  0.0149 *
direct:f.type.centre3  2.7083❼  1.1012    2.459  0.0139 *
---
Signif. codes: 0 '***' 0.001 '**' 0.01 '*' 0.05 '.' 0.1 ' ' 1
```

```
(Dispersion parameter for binomial family taken to be 1)

    Null deviance: 659.51 on 663 degrees of freedom
Residual deviance: 370.15 on 650 degrees of freedom
  (135 observations deleted due to missingness)
AIC: 398.15

Number of Fisher Scoring iterations: 6
        ❽
> drop1(mod, .~., test = "Chisq")
Single term deletions

Model:
suicide.hr ~ grav.cons + suicide.past + dep.cons + past.prison
    + sczdys.cons + death + against.people + subst.cons + direct
    + f.type.centre + direct:f.type.centre
               Df Deviance    AIC    LRT   Pr(Chi)
<none>             370.15  398.15
grav.cons       1  405.54  431.54 35.38 2.706e-09 ***
suicide.past    1  413.24  439.24 43.09 5.220e-11 ***
dep.cons        1  396.28  422.28 26.13 3.192e-07 ***
past.prison     1  376.80  402.80  6.65 0.00992   **
sczdys.cons     1  373.15  399.15  3.00 0.08343   .
```

```
death                     1    371.25   397.25   1.10  0.29523
against.people            1    370.78   396.78   0.63  0.42900
subst.cons                1    370.76   396.76   0.61  0.43528
direct                    1    371.09   397.09   0.94  0.33183    ❾
f.type.centre             2    371.49   395.49   1.34  0.51282
direct:f.type.centre      2    378.03   402.03   7.88  0.01947   *❿
---
Signif. codes:  0 '***' 0.001 '**' 0.01 '*' 0.05 '.' 0.1 ' ' 1
```

It is first necessary to verify the coding of the two variables present in the interaction term; "direct"❶ and "f.type.centre"❷ are coded here as (1,0). Because "f.type.centre" is a categorical variable with three levels, it is recoded with two binary variables❷. In the function glm(), the interaction term "direct × f.type.centre" is represented by direct:f.type.centre❹. The global effect of each variable is obtained from the instruction drop1()❽. The interaction "direct × f.type.centre" is statistically significant❿; the main effect "direct" is no longer significant❾ but this should be considered very cautiously (and perhaps not taken into consideration at all) because this is a main effect in the presence of an interaction term that contains it. Always about variable "direct," how can the coefficient ❺ be interpreted? As explained above, the associated odds-ratio equal to exp(−1.45❺) = 0.24 is related to the sub-population of prisoners for which f.type.centre2 = 0 and f.type.centre3 = 0. From ❸ this corresponds to type 1 prisons 1 (high-security units): a low level of self-directedness is associated with a risk of suicide attempt divided by 4.17 (1/4.17 = 0.24) in high-security units (we had hypothesised the reverse). However, this association is not statistically significant ($p = 0.16$). For prisoners in type 2 prisons (centres for longer sentences or for prisoners with good scope for rehabilitation), from ❸ we note that f.type.centre2 = 1 and f.type.centre3 = 0 so that the odds-ratio is now exp(−1.45❺ + 2.85❻) = 4.1. Here, a low level of self-directedness is a risk factor for suicide attempt. For prisoners in type 3 prisons, we have exp(−1.45❺ + 2.71❼) = 3.5. To test whether or not these last two odds-ratios are statistically different from 1, we can use the function estimable() in the package "gmodels" (see Section 6.2):

```
> library(gmodels)
> estimable(mod, c(0, 0, 0, 0, 0, 0, 0, 0, 0, 1, 0, 0, 1, 0),
  conf.int = 0.95)
                              Estimate    Std. Error X^2 value DF
(0 0 0 0 0 0 0 0 0 1 0 0 1 0)  1.402069    0.5368913  6.819711  1
                              Pr(>|X^2|)   Lower.CI   Upper.CI
(0 0 0 0 0 0 0 0 0 1 0 0 1 0) ❿0.009015716 0.3368918  2.467246
> estimable(mod, c(0, 0, 0, 0, 0, 0, 0, 0, 0, 1, 0, 0, 0, 1),
  conf.int = 0.95)
```

```
                          Estimate   Std. Error X^2 value DF
(0 0 0 0 0 0 0 0 1 0 0 0 1)   1.261613   0.3877241  10.58782  1
                          Pr(>|X^2|)   Lower.CI   Upper.CI
(0 0 0 0 0 0 0 0 1 0 0 0 1) ❷0.001138351 0.4923792  2.030846
```

Both p-values are lower than 0.05 in ❶ and ❷.

Now if we use a (1,–1) coding for the variables "direct" and "f.type.prison," we obtain

❶

```
> mhp.mod$direct2 <- ifelse(mhp.mod$direct == 1, 1, -1)
  ❷
> table(mhp.mod$direct, mhp.mod$direct2, useNA = "ifany")

         -1    1   <NA>
0        567    0     0
1          0  147     0
<NA>       0    0    85
```

❸

```
> contrasts(mhp.mod$f.type.centre) <- contr.sum
> contrasts(mhp.mod$f.type.centre)
     [, 1]   [, 2]
1      1      0
2      0      1
3     -1     -1
> mod <- glm(suicide.hr ~ grav.cons + suicide.past
  + dep.cons + past.prison + sczdys.cons + death
  + against.people + subst.cons + direct2 + f.type.centre
  + direct2:f.type.centre, data = mhp.mod, family = "binomial")
> summary(mod)

Call:
glm(formula = suicide.hr ~ grav.cons + suicide.past + dep.cons
    + past.prison + sczdys.cons + death + against.people
    + subst.cons + direct2 + f.type.centre + direct2:f.type.centre,
    family = "binomial", data = mhp.mod)

Deviance Residuals:
   Min       1Q     Median      3Q        Max
-2.03462  -0.43702  -0.19137  -0.07424   3.16051


Coefficients:
                Estimate  Std. Error  z value  Pr(>|z|)
(Intercept)      -6.1500     0.6633    -9.272   < 2e-16 ***
grav.cons         0.6692     0.1207     5.545  2.94e-08 ***
suicide.past      1.7729     0.2791     6.352  2.13e-10 ***
```

```
dep.cons                     1.4319     0.2893      4.950  7.43e-07 ***
past.prison                  0.7321     0.2867      2.554  0.0107 *
sczdys.cons                  1.3722     0.8377      1.638  0.1014
death                        0.3008     0.2868      1.049  0.2943
against.people              -0.2359     0.2984     -0.791  0.4292
subst.cons                  -0.2405     0.3100     -0.776  0.4380
direct2                      0.2028❹    0.2031      0.999  0.3179
f.type.centre1              -0.6359     0.3681     -1.728  0.0840 .
f.type.centre2               0.2242     0.2617      0.857  0.3916
direct2:f.type.centre1      -0.9262     0.3623     -2.557  0.0106 *
direct2:f.type.centre2       0.4982     0.2584      1.928  0.0538 .
---
Signif. codes:  0 '***' 0.001 '**' 0.01 '*' 0.05 '.' 0.1 ' ' 1

(Dispersion parameter for binomial family taken to be 1)

    Null deviance: 659.51 on 663 degrees of freedom
Residual deviance: 370.15 on 650 degrees of freedom
  (135 observations deleted due to missingness)
AIC: 398.15

Number of Fisher Scoring iterations: 6

> drop1(mod, .~., test = "Chisq")
Single term deletions

Model:
suicide.hr ~ grav.cons + suicide.past + dep.cons + past.prison
   + sczdys.cons + death + against.people + subst.cons
   + direct2 + f.type.centre + direct2:f.type.centre
                      Df Deviance    AIC    LRT    Pr(Chi)
<none>                    370.15  398.15
grav.cons              1  405.54  431.54  35.38  2.706e-09 ***
suicide.past           1  413.24  439.24  43.09  5.220e-11 ***
dep.cons               1  396.28  422.28  26.13  3.192e-07 ***
past.prison            1  376.80  402.80   6.65  0.00992 **
sczdys.cons            1  373.15  399.15   3.00  0.08343 .
death                  1  371.25  397.25   1.10  0.29523
against.people         1  370.78  396.78   0.63  0.42900
subst.cons             1  370.76  396.76   0.61  0.43528
direct2                1  371.09  397.09   0.94  0.33183    ❺
f.type.centre          2  374.14  398.14   3.99  0.13605    ❻
direct2:f.type.centre  2  378.03  402.03   7.88  0.01947 *❼
---
Signif. codes:  0 '***' 0.001 '**' 0.01 '*' 0.05 '.' 0.1 ' ' 1
```

The variable "direct" is recoded in ❶, and the modification is verified in ❷. The variable "f.type.centre" is recoded in ❸ with the "contr.sum" option of the contrasts() function (which could also have been used with "direct" if it had been defined as a factor instead of a binary numerical variable). The drop1() function leads to similar results for the interaction term ❼ with the two codings (and this is expected), but gives different estimates of p-values for the main effect ❻. The coefficient of "direct" is equal to 0.2028❹. It can be interpreted as follows: Because the coding is (1,–1), the odds-ratio associated with this variable is exp(2 × 0.2028) = 1.5. This odds-ratio involves the effect of low self-directedness on being at high risk of suicide attempt, in a population where the three types of prison are assumed to be balanced, and given that the other predictor variables in the model are held constant.

Now it is time to have a look at the procedure proposed by Cox and Wermuth (1996). This procedure gives an idea of the overall importance of interaction terms in a given model. First, all interaction terms are estimated and tested, one at a time. Second, the corresponding p-values are plotted in a diagram, and a visual assessment can help determine if there is indeed a substantial amount of interaction terms that should be retained. It would be tedious to perform all these computations using a classical statistical package. With R, it takes only a few lines:

```
              ❶
> mod <- glm(suicide.hr ~ grav.cons + suicide.past + dep.cons
    + direct + past.prison + sczdys.cons + death + against.people
    + subst.cons + f.type.centre, data = mhp.mod,
    family = "binomial")
                      ❷            ❸            ❹
> mod.inter <- data.frame(add1(mod, ~.^2, test = "Chisq"))
            ❺                ❻
> isort <- order(mod.inter["Pr.Chi."])
              ❼
> mod.inter[isort, ]
                          Df Deviance    AIC       LRT        Pr.Chi.
past.prison:against.people 1 370.8583 396.8583 7.171484133 0.007407147❽
suicide.past:dep.cons      1 371.6522 397.6522 6.377570255 0.011557154
direct:f.type.centre       2 370.1516 398.1516 7.878152389 0.019466189
dep.cons:sczdys.cons       1 373.7124 399.7124 4.317342548 0.037725795
suicide.past:direct        1 374.5811 400.5811 3.448631713 0.063304155
grav.cons:against.people   1 375.2116 401.2116 2.818140682 0.093204312
dep.cons:against.people    1 375.3861 401.3861 2.643630152 0.103965739
grav.cons:dep.cons         1 375.5737 401.5737 2.456018505 0.117075207
sczdys.cons:f.type.centre  2 373.7939 401.7939 4.235870649 0.120279711
against.people:subst.cons  1 375.9531 401.9531 2.076679837 0.149565193
past.prison:f.type.centre  2 374.2442 402.2442 3.785512614 0.150655982

.... [TRUNCATED]
```

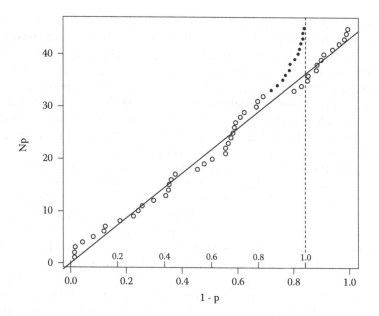

FIGURE 6.3
Graphical representation of the p-values of the 45 possible interaction terms in a logistic regression model. Under the assumption that all interaction terms are null, the empty dots should be on a straight line (each dot corresponds to an interaction). This is observed in the present situation. The solid dots correspond to an imaginary situation where there seems to be surprisingly small p-values for some interaction terms.

The basic model without any interaction is in ❶. The function add1()❸ is then called to add all interaction terms as specified in ❹, one at a time. The corresponding results are stored in a data frame❷. The index of the line with the smallest p-value❻, the next smallest, etc. is obtained using the function order(). The sorted results are presented in ❼; the interaction term with the smallest p-value is in ❽.

These p-values can be plotted (Figure 6.3):

```
              ❶
> x <- 1-na.omit(mod.inter[isort, "Pr.Chi."])
              ❷
> y <- length(x):1
> plot(x, y, xlab = "1 - p", ylab = "Np")
              ❸                        ❹
>       line <- lm(y[length(x):1] ~ -1 + x[length(x):1])
       ❺
> abline(line)
```

In this plot (Figure 6.3), the ordered p-values❶ obtained previously are represented on the x-axis, and the numbers of the interaction terms❷ are represented on the y-axis, ranked from the largest p-value to the smallest. Under the assumption that all interaction terms are null, the distribution of the p-values should be uniform and the points on the diagram should be on a line passing through the origin. For convenience, this line is represented (function abline()❺). It is obtained from the function lm()❸ and a specification that the line passes through the origin (~ -1+❹). It can be noted that in the present example, the empty dots are approximately on the line: the problem of interaction is globally negligible. The full dots correspond to an imaginary situation where there seems to be surprisingly small p-values for some interaction terms. These interaction terms should be investigated carefully.

What should be concluded from all these analyses? If the interaction between "direct" and "f.type.centre" had actually been anticipated before the analysis, then the results are of interest: It appears that a low level of self-directedness globally increases the probability of being at high risk of suicide attempt (this is derived from the analysis without the interaction term); it also appears that this effect is small (and perhaps reversed) in high-security units, while it is more marked in the other types of prison. This result is the reverse of what was expected, and is therefore difficult to interpret. Perhaps life is particularly ritualized in high-security units so that the need for strong self-directedness could be less crucial in enduring imprisonment. In all events, this surprising result shows that it is difficult to predict an interaction between two variables, and that explanations can always be proposed *a posteriori*.

6.5 Assessing the Relative Importance of Predictors

In a few words: The ultimate aim of a model is often to determine which predictors are likely to have a stronger impact on the outcome. Because this determination is crucial, there is a large volume of literature on it. Unfortunately, this literature is particularly controversial (O'Grady 1982; Greenland, Schlesselman, and Criqui 1986; Greenland et al. 1991; Rothman and Greenland 1998; Menard 2004; Mood 2009).

A possible explanation of these controversies is that the notion of "strong impact" is polysemous and that its formal translation can vary from one discipline to another.

In epidemiology (for example), most models are logistic regressions that explain a binary variable (having a disease or not) by a series of binary risk factors. In this case, the exponential of the regression coefficients are adjusted odds-ratios (*aOR*), and they can be used to represent the impact of a given risk factor on the probability of having the disease. Indeed, when the prevalence of the disease is low, the probability of having the disease is approximately *aOR* times greater when the subjects have the risk factor than when they do not (assuming the other predictor variables in the model are held constant). An odds-ratio has a clear clinical meaning, but this does not mean that it summarizes all aspects of the question of the "importance" of a risk factor. For instance, if a risk factor is associated with a large aOR but involves only a very small proportion of the population, then this risk factor is perhaps not so "important" (at least from a public health perspective).

To tackle this point, epidemiologists have defined the notion of the "attributable fraction" (AF), which is defined as the percentage of cases (i.e., subjects presenting the disease) that would have not occurred if exposure to the risk factor had not occurred (Rothman and Greenland 1998). The AF can be easily estimated from the aOR (or, better, the adjusted relative risk) and from the percentage p of cases that are exposed to the risk factor:

$$AF = p\frac{aOR - 1}{aOR}$$

If we now consider linear models, regression coefficients can also give an indication of the impact that a predictor could have on the outcome. For instance, suppose we have $Y = a_0 + a_1 \times X_1 + \ldots + a_p \times X_p + \varepsilon$, with Y being the "income per year in €" and X_1 the variable "age" (in years). If $a_1 = 200$, then, on average, a subject's income increases by 200€ each year (assuming the other predictor variables in the model are held constant).

Now, if X_2 is "gender (1 for male, 0 for female)" and if $a_2 = 10{,}000$, is it possible to compare the magnitude of the effects of age and gender on the income per year from the magnitudes of a_1 and a_2? In one sense it is, because from a_1 and a_2 it is possible to say that there is a difference of about 10,000€ between men and women and a difference of about 4,000€ between a person aged 30 and another aged 50. Thus, the gender effect appears stronger compared to age. However, one could claim that it is meaningless to compare a_1 and a_2, as age and gender have nothing in common and, in particular, their units and their scales are totally different.

To deal with this issue, it is usual in certain fields to rescale the variables before the model is estimated so that they are all on comparable footing. In practice, this rescaling consists of converting the variables in a linear manner so that they have a mean equal to 0 and a variance equal to 1. In this situation, the variables are said to be "standardized," and the regression coefficients are called "standardized regression coefficients." This approach can be useful in questionnaire surveys when the answers are different from one question to the other, with no evocative unit (e.g., Q1 (satisfied/not satisfied), Q2 (not satisfied at all/not satisfied/rather satisfied/satisfied/very satisfied), Q3 (rarely satisfied/sometimes satisfied/often satisfied), etc.). This approach, however, should not be used systematically because it may be that the wordings of response choices *are* meaningful, and because standardization does not settle all the problems (Greenland, Schlesselman, and Criqui 1986; Greenland et al. 1991).

Two arguments are often opposed to the standardization of coefficients. First, two studies dealing with the same phenomenon and the same variables, but in populations where the variances of these variables are different, will lead to different standardized regression coefficients (but to similar raw regression coefficients). In other words, standardized regression coefficients are not simply characteristics of the phenomenon under study; they are also functions of the sample that is observed. Second, it may be easy to modify a given predictor (e.g., consumption of milk) while this will be difficult and even impossible for another predictor (age, for instance). If the underlying objective of a study is to plan an intervention based on "important" predictors, then standardized coefficients can only be of limited interest to characterise this "importance."

Note that measures of explained variance (see below and in Section 3.2 for details), which are also greatly favoured in some disciplines, have potentially comparable advantages and drawbacks (O'Grady 1982).

Finally, economists have developed their own tool to assess the relative weight of predictors in a model: elasticity. Elasticity is the ratio of a percentage of change in the outcome to a percentage of change in the predictor. It is particularly well-suited to variables for which percentages are meaningful, prices being a good example.

In Practice: In the previous sections we looked at logistic regression models that explained the outcome variable "being at high risk for suicide attempt (yes/no)" by a series of predictors, among which being depressed (yes/no). The odds-ratio associated with this predictor was exp(1.36) = 3.9 in a stepwise selection model (see Section 6.3). Certain other predictor odds-ratios were even greater than that for depression, but the prevalence of depression is high and there exist treatments to cure this disorder (as compared to "history of suicide attempt," for example, which is irretrievable). It could therefore be useful to estimate the attributable fraction associated with depression in "being at high risk of suicide attempt." For this estimation, we need: (1) the prevalence of depression in the population of prisoners at high risk of suicide attempt and (2) the adjusted relative risk odds-ratio for depression. For this purpose, the following calculations are used:

```
> mhp.mod$pw[mhp.mod$type.centre2 == 1] <- 20
> mhp.mod$pw[mhp.mod$type.centre2 == 2] <- 45
> mhp.mod$pw[mhp.mod$type.centre2 == 3] <- 67
> mhp.mod$strat.size[mhp.mod$type.centre2 == 1] <- 13
> mhp.mod$strat.size[mhp.mod$type.centre2 == 2] <- 55
> mhp.mod$strat.size[mhp.mod$type.centre2 == 3] <- 144
                     ❶
> library(survey)
                                                      ❷
> mhp.modhr <- mhp.mod[mhp.mod$suicide.hr == 1 &! is.na(mhp.mod$
  suicide.hr), ]
```

```
                         ❸
> table(mhp.modhr$suicide.hr)
  1
153
> mhp.surveyhr <- svydesign(id = ~centre,
  strata = ~type.centre2, weights = ~pw, fpc = ~strat.size,
  data = mhp.modhr)
                ❹
> res1 <- svymean(~dep.cons, mhp.surveyhr, na.rm = TRUE)
  ❺
> p <- res1[1]
> mhp.survey <- svydesign(id = ~centre, strata = ~type.centre2,
  weights = ~pw, fpc = ~strat.size, data = mhp.mod)
            ❻
> res <- svyglm(suicide.hr ~ dep.cons + grav.cons
  + suicide.past + past.prison + sczdys.cons + death
  + against.people + subst.cons + direct + f.type.centre,
  design = mhp.survey, na.action = na.omit, family = "binomial")
Warning message:
In eval(expr, envir, enclos) : non-integer #successes in a
  binomial glm!
        ❼
> aOR <- exp(res$coefficients["dep.cons"])
          ❽
> AF <- p*(aOR - 1) / aOR
> AF
  dep.cons
0.6310984
```

Because the sampling design of the MHP study is a two-level random sampling with unequal weights, we need to estimate the prevalence of depression in the population of prisoners at high risk of suicide attempt using the "survey" package❶. We select prisoners at risk in ❷ and note the corresponding sample size in ❸. The estimate of the prevalence of depression ❺ is then obtained in this subsample using the function svymean()❹.

The adjusted odds-ratio (*aOR*) is also estimated using the "survey" package and the svyglm() function❻; it is equal to the exponential of the coefficient of the variable depression❼. The attributable fraction is computed in ❽ and gives 63%. This result should be considered with extreme caution. First, the outcome is not a rare event (about 20% of prisoners present a high risk for suicide attempt) so that the *aOR* is not a good estimate of the adjusted relative risk that is theoretically required in the estimation of the attributable fraction. Second, the attributable fraction computation is based on the hypothesis that there is a causal process between depression and risk of suicide attempt. The logistic regression model is an attempt to deal with this issue (finding the role of depression assuming the other predictor variables in the model are held constant), but there is indeed no guarantee that this has been successful. Nevertheless, what we do have here is actually interesting from a public health perspective because, in a French prison, depression *might* explain more than half of the cases of prisoners at high risk of suicide attempt.

If we now look for an example of linear regression, let us consider the model that explained the duration of interview in minutes (see Section 5.1). In the following output we estimate successively (1) a linear regression model based on raw data, (2) the standardized variables using the function scale(), and (3) a linear regression model based on these standardized variables:

❶

```
> mod <- lm(dur.interv ~ scz.cons + dep.cons + subst.cons
  + grav.cons + char + trauma + age, data = mhp.mod2)
> summary(mod)

Call:
lm(formula = dur.interv ~ scz.cons + dep.cons + subst.cons
  + grav.cons + char + trauma + age, data = mhp.mod2)

Residuals:
   Min       1Q    Median       3Q      Max
-39.2296  -14.0525  -0.7474  10.8543  63.2329

Coefficients:
             Estimate Std. Error t value Pr(>|z|)
(Intercept) 42.01471    3.18775  13.180  < 2e-16 ***
scz.cons     3.05140❷   2.80978   1.086  0.277888
dep.cons     6.47369❸   1.65515   3.911  0.000102 ***
subst.cons   4.31697    1.79807   2.401  0.016636 *
grav.cons    0.73050    0.55949   1.306  0.192131
char         2.04442    0.93201   2.194  0.028621 *
trauma      -1.02904    1.67282  -0.615  0.538670
age          0.25831❹   0.05866   4.404  1.24e-05 ***
---
Signif. codes:  0 '***' 0.001 '**' 0.01 '*' 0.05 '.' 0.1 ' ' 1

Residual standard error: 18.14 on 647 degrees of freedom
  (91 observations deleted due to missingness)
Multiple R-squared: 0.09397, Adjusted R-squared: 0.08417
F-statistic: 9.587 on 7 and 647 DF, p-value: 2.287e-11
```

❺

```
> mhp.mod3 <- data.frame(scale(mhp.mod2[, c("dur.interv",
  "scz.cons", "dep.cons", "subst.cons", "grav.cons", "char",
  "trauma", "age")]))
> library(prettyR)
```

```
> describe(mhp.mod3)
Description of mhp.mod3

Numeric            ❻                    ❼
                mean      median  var  sd  valid.n
dur.interv   -3.305e-17  -0.1091   1    1    746
scz.cons      2.906e-17  -0.2982   1    1    746
dep.cons      4.17e-17   -0.788    1    1    746
subst.cons   -2.001e-17  -0.5883   1    1    746
grav.cons     9.096e-17   0.2293   1    1    742
char          1.369e-16  -0.6028   1    1    662
trauma        4.621e-17  -0.5911   1    1    737
age           2.065e-16  -0.0853   1    1    744

> mod <- lm(dur.interv ~ scz.cons + dep.cons + subst.cons
   + grav.cons + char + trauma + age, data = mhp.mod3)
> summary(mod)

Call:
lm(formula = dur.interv ~ scz.cons + dep.cons + subst.cons
   + grav.cons + char + trauma + age, data = mhp.mod3)

Residuals:
    Min       1Q      Median      3Q        Max
-2.02363  -0.72489  -0.03855   0.55991   3.26181

Coefficients:
              Estimate   Std. Error  t value  Pr(>|t|)
(Intercept)  -0.03402    0.03664     -0.928   0.353582
scz.cons      0.04316    0.03974      1.086   0.277888
dep.cons      0.16247❽   0.04154      3.911   0.000102  ***
subst.cons    0.09742    0.04058      2.401   0.016636  *
grav.cons     0.06134    0.04698      1.306   0.192131
char          0.09091    0.04145      2.194   0.028621  *
trauma       -0.02327    0.03784     -0.615   0.538670
age           0.17742❾   0.04029      4.404   1.24e-05  ***
---
Signif. codes:  0 '***' 0.001 '**' 0.01 '*' 0.05 '.' 0.1 ' ' 1

Residual standard error: 0.9359 on 647 degrees of freedom
  (91 observations deleted due to missingness)
Multiple R-squared: 0.09397, Adjusted R-squared: 0.08417
F-statistic: 9.587 on 7 and 647 DF, p-value: 2.287e-11
```

The classic linear regression model is estimated in ❶. Assuming the other predictor variables in the model are held constant, the impact of schizophrenia on duration of interview is about 3 minutes❷ (but this is not statistically significant) and about six minutes for depression❸. Being older by one year leads to a duration of interview that increases by a quarter of a minute (or, equivalently, being older by 10 years leads to an increase of four minutes)❹.

The function scale() is used in ❺ to standardize the variables. This is verified using the function describe() (package "prettyR"): Means are equal to 0❻ and variance to 1❼. A new call to the function lm() now gives the standardized coefficients. An increase of about 1/6 of a standard deviation in duration of interview can be observed from an increase of 1 standard deviation for the variable "dep.cons"❽, and the same is true for "age"❾.

Let us see now how to estimate the amount of variance of the outcome "duration of interview" that is explained by each predictor. The function calc.relmp() from the package "relaimpo" (Grömping 2006) is specifically dedicated to this task:

```
> library(relaimpo)
                              ❶          ❷
> calc.relimp(mod, type = c("lmg", "last"), rela = TRUE)
Response variable: dur.interv
Total response variance: 0.9563792
Analysis based on 655 observations

7 Regressors:
scz.cons dep.cons subst.cons grav.cons char trauma age
Proportion of variance explained by model: 9.4%
Metrics are normalized to sum to 100% (rela=TRUE).

Relative importance metrics:

                  lmg                last
scz.cons     0.040472107        0.024302114
dep.cons     0.344825510❸       0.315227494
subst.cons   0.070825564        0.118778441
grav.cons    0.167247794        0.035128094
char         0.151931790        0.099150492
trauma       0.004625009        0.007797534
age          0.220072224❹       0.399615831
.... [TRUNCATED]
```

Measuring explained variance is easy when the predictors are independent variables because, in this situation, the total variance can be divided straightforwardly into shares for each predictor. Unfortunately, in practice, predictors are not independent, so that it is difficult to define the share of the outcome variance that is specifically related to each of them. Classically, these shares of variance correspond to the increase in R2 if a given predictor is added to the model that already includes all the other predictors (Grömping 2007). This approach is, however, not optimal because the sum of all shares of variance does not correspond to the total variance. Nevertheless, because they are classic, the increases in R2 are requested in ❷. The "Lindeman, Merenda, and Gold" (lmg) approach is currently preferred (Grömping 2007) and is requested in ❶. The results indicate that about 34% of the outcome variance is explained by depression ❸ and about 22% by age ❹. These figures must be interpreted with caution, in particular because the notion of "explaining a certain % of the outcome variance" is first and foremost formal and not so intuitive.

6.6 Dealing with Missing Data

In a few words: Missing data are a major concern for researchers engaged in questionnaire surveys. There is no doubt that the best approach with regard to this issue is prevention. Indeed, if questions are carefully drafted and tested in a pilot study, and if interviewers are well trained and not too hurried, then the quality of information will improve and missing data will be the exception rather than the rule. Nevertheless, even with the greatest care, there *will* be missing data, and the question of how to deal with this therefore must be raised in the course of the analysis.

Basically, two approaches can be envisaged to answer this question.

The first consists of discarding the subjects who miss one or several responses that are necessary to carry out the analysis. The advantage of this approach is its simplicity. It, however, has several drawbacks, including

1. The sample under consideration is likely to vary according to the pattern of variables involved in each analysis; this can lead to statistical problems (see Section 6.3).

2. An incomplete observation is rarely totally incomplete, so that dropping incomplete subjects leads to a loss of information and a loss of statistical power.

3. If non-responders have a particular profile (e.g., people who do not answer questions concerning alcohol consumption are likely to be heavier drinkers than the others), then the results of the analyses can be biased (Harrel 2001).

The second approach to deal with missing values consists of replacing them by artificial values; this is called "imputation." The advantages here are that there is no loss of power and that the sample analysed remains the same. The main drawback is that the sample is composed of a mix of real data and data created artificially by the investigator; this can appear as not very scientific or even somewhat dishonest.

Should we delete or impute? In practice, the choice depends more on disciplinary habit than on statistical considerations. In epidemiology, imputations were not viewed kindly for a long time. In social sciences, when a principal component analysis (Section 3.6) or a factor analysis (Section 7.2) is run on a large questionnaire, imputations are necessary because the percentage of complete observations can be dramatically small.

A frequent argument against imputation is that when a single question with important repercussions is to be answered on the basis of the results of a statistical test of hypothesis, a type 1 error is no longer valid if fake (imputed) data are added to real data. Indeed, a p-value is related to "what could have been observed by chance" (see Section 4.4) and there seems to be no "chance" in the constitution of fake data. This is, nevertheless, only partially true. First, if the imputations concern only adjustment variables in a regression model and not the outcome or the main predictor of interest, then the concomitant artificial increase in sample size and the systematic underestimation of p-values will likely be less problematic. Second, there are imputation techniques that incorporate "chance" into the imputation procedure, so that the statistical tests do not present a systematic underestimation of the p-values. The principle of these techniques is to perform several imputations (e.g., five) with a certain amount of noise added to the values imputed. The analyses are then performed on the five imputed datasets and five series of results are obtained. Finally, a global solution is produced that takes into account the variation due to the estimation process in each model and the variation observed between the five datasets. This is called a "multiple imputation" process and is considered to have appreciable statistical properties (Harrel 2001; Van Buuren et al. 2006). Of course, multiple imputation techniques are somewhat cumbersome to implement. Fortunately, some R packages offer efficient routines that are straightforward to use with a linear, logistic, or Poisson regression models.

To summarize:

- The problem of missing data should be guarded against as far as possible when designing the study and the questionnaire.
- Nevertheless, because there are always missing data, the issue should be explicitly addressed at the statistical analysis stage.
- As usual, the first step in the analysis is descriptive. What is the overall extent of the problem? Which variables are most often involved? Are there explanations (e.g., subjects could be embarrassed by the question or not feel concerned by it, etc.)?

- Deletion of subjects or imputation of missing data are then the two alternatives. In practice, most researchers delete because it is easier; from a statistical point of view; this may not be the best approach. Multiple imputation techniques are potentially valuable, especially when there are numerous observations with missing data.

- General usage in the particular discipline must also be considered.

In Practice: The functions `summary()` and `describe()` (package "prettyR") provide, for each variable, both the number of missing observations and the most basic descriptive statistics.

It happens frequently that the same group of variables is missing for a given sub-sample of subjects (e.g., people can refuse to answer all questions related to religion or to political opinions). The function `naclus()` in the package "Hmisc" is designed to find clusters of missing data of this type; it is informative and simple to use:

```
> plot(naclus(mhp.mod))
```

The output is a hierarchical clustering (see Section 3.5) (Figure 6.4). In ❶, we note that the two variables "n.prison" (number of past imprisonment) and "d.prison" (duration of past imprisonment) have about 50% of missing observations with a close pattern of non-response. This is due to the fact that interviewers asked these two questions only if the answer to the question on "past imprisonment" was positive. In ❷, we can see that all "character" variables in the Cloninger character and temperament scale have a similar pattern of non-response; the same is true in ❸ for "temperament" variables. In ❹ and ❺, there are two other small clusters, one related to suicide and the other to the duration of interview.

Let us imagine now that the particular research focus is the character variable "direct" (low level of self-directedness: yes[1]/no[0]). In Figure 6.4 it appears that this variable presents at least 10% missing observations so that if "direct" is a predictor in a model, about 10% of the observations collected in the MHP study must be discarded and this is problematic.

Before dealing with the question of deleting or imputing observations with missing data for the variable "direct," we need to determine if the corresponding prisoners have particular characteristics. Because in Figure 6.4 we can see in ❷ that when a subject has missing "direct" values, he most often has missing values for all the dimensions of the Cloninger temperament and character scale; it may be preferable to focus on the whole Cloninger instrument rather than on a particular dimension. A focused principal component analysis (function `fpca()` in the package "psy"; see Section 3.8) can be useful for this:

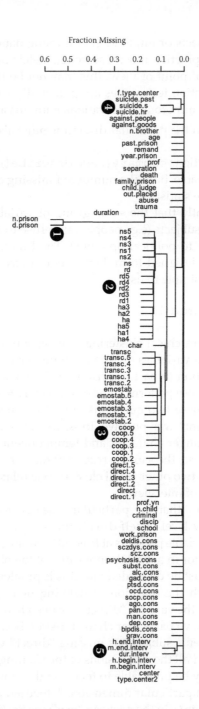

FIGURE 6.4
Hierarchical clustering of variables according to the similarity of their patterns of missing observations. In ❷, all "character" variables have missing values for the same subset of prisoners. The same is true in ❸ for the "temperament" variables.

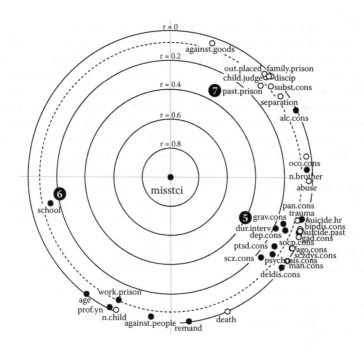

FIGURE 6.5

Focused principal component analysis used to determine variables associated with "misstci" (having at least one missing value in the Cloninger instrument). Variables inside the dashed line are significantly associated with "misstci." Black dots correspond to positive associations, empty dots to negative associations.

❶

```
> mhp.mod$misstci <- is.na(mhp.mod$direct) | is.na(mhp.mod$coop)
  | is.na(mhp.mod$emostab) | is.na(mhp.mod$transc)
  | is.na(mhp.mod$ha) | is.na(mhp.mod$rd) | is.na(mhp.mod$ns)
> library(psy)
```

❷   **❸**

```
> fpca(misstci ~ suicide.hr + grav.cons + bipdis.cons + dep.cons
  + man.cons + pan.cons + ago.cons + socp.cons + ocd.cons
  + ptsd.cons + gad.cons + alc.cons + subst.cons
  + psychosis.cons + scz.cons + sczdys.cons + deldis.cons
  + n.child + n.brother + school + prof.yn + remand
  + against.goods + against.people + work.prison + discip
  + past.prison + child.judge + out.placed + death + separation
  + abuse + trauma + family.prison + age + suicide.past + age
  + dur.interv, data = mhp.mod, contraction = "Yes"❹)
```

The new variable "misstci" is "TRUE" if at least one of the seven dimensions in the Cloninger instrument is missing❶ and is otherwise "FALSE." The relationships between "misstci"❷ and 37 variables ❸ (numerical or binary) are represented graphically in Figure 6.5. One cluster of variables ❺

(Figure 6.5) is significantly associated with "misstci"; it contains "dur.interv" (duration of interview), "grav.cons" (consensual assessment of gravity), "dep. cons" (consensual diagnosis of depression), and other diagnoses. This can be interpreted easily. The Cloninger instrument is administered at the end of the interview. If the interview lasts too long because the prisoner presents one or more mental disorders, then there is an increased probability that the interview will terminate prematurely due do to external constraints (security, meeting with a lawyer, time for dinner, etc.). In this situation, the last variables will not be collected, in particular the seven dimensions of the Cloninger instrument. Having a high-school grade ❻ (Figure 6.5) or no past imprisonment ❼ (Figure 6.5) are also positively associated with a high probability of missing data on the Cloninger scale. The interpretation is less clear here; perhaps we have two indicators for people who are less inclined to answer questions in this particular context.

Another way to study the characteristics of prisoners who have missing values in the Cloninger scale is to use a logistic model with "misstci" as outcome:

```
> mod <- glm(misstci ~ school + past.prison + dur.interv
  + grav.cons + dep.cons + scz.cons + psychosis.cons
  + ptsd.cons + f.type.centre, data = mhp.mod,
  family = "binomial")
> summary(mod)

Call:
glm(formula = misstci ~ school + past.prison + dur.interv
  + grav.cons + dep.cons + scz.cons + psychosis.cons
  + ptsd.cons + f.type.centre, family = "binomial",
  data = mhp.mod)

Deviance Residuals:
  Min       1Q      Median      3Q       Max
-1.5576   -0.6942   -0.4894   -0.3036   2.6738
```

Coefficients:

	Estimate	Std. Error	z value	Pr(>|z|)	
(Intercept)	-3.125443	0.509789	-6.131	8.74e-10	***
school	0.342587	0.094225	3.636	0.000277	***❶
past.prison	-0.614984	0.202946	-3.030	0.002443	** ❷
dur.interv	0.024091	0.004915	4.902	9.50e-07	***❸
grav.cons	0.028718	0.082517	0.348	0.727820	
dep.cons	0.254821	0.218377	1.167	0.243256	
scz.cons	1.111866	0.359340	3.094	0.001974	** ❹
psychosis.cons	-0.196543	0.293088	-0.671	0.502480	
ptsd.cons	0.461153	0.233175	1.978	0.047961	*
f.type.centre2	-1.261115	0.325681	-3.872	0.000108	***❺
f.type.centre3	-0.352828	0.273269	-1.291	0.196656	

```
---
```

```
Signif. codes: 0 '***' 0.001 '**' 0.01 '*' 0.05 '.' 0.1 ' ' 1

(Dispersion parameter for binomial family taken to be 1)

    Null deviance: 763.44 on 736 degrees of freedom
Residual deviance: 668.78 on 726 degrees of freedom
 (62 observations deleted due to missingness)
AIC: 690.78

Number of Fisher Scoring iterations: 5
```

A high-school grade❶, no history of past imprisonment❷, a long duration of interview❸, presenting a schizophrenic disorder❹, and not being in a type 2 prison❺ are understandably among the most salient predictors.

Now, if we return to the problem of missing values involving the "direct" character variable, what are the options if it is decided not to delete the corresponding observations?

An apparently ingenious way of dealing with this issue is sometimes proposed; it consists of recoding "direct" as a categorical variable with three levels: Yes[1]/No[0]/missing[missing]:

```
                         ❶
> mhp.mod$direct.miss <- factor(mhp.mod$direct)
                                               ❷
> levels(mhp.mod$direct.miss) <- c("0", "1", "missing")
                               ❸
> mhp.mod$direct.miss[is.na(mhp.mod$direct.miss)] <- "missing"
      ❹
> table(mhp.mod$direct.miss, mhp.mod$direct, useNA = "ifany")

                 0     1   <NA>
  0            567     0      0
  1              0   147      0
  missing        0     0     85
```

The new variable "direct.miss" is defined in ❶ from the "direct" variable. The three levels are identified in ❷. All "NA"❸ are replaced by "missing." In ❹, it is verified that the recoding works satisfactorily.

If this variable is now used in a regression model, there are two coefficients for "direct.miss" instead of one, as was the case for "direct":

```
> mod <- glm(suicide.hr ~ dep.cons + grav.cons + suicide.past
  + past.prison + sczdys.cons + death + against.people
  + subst.cons + direct.miss + f.type.centre, data = mhp.mod,
  family = "binomial")
> summary(mod)
```

```
Call:
glm(formula = suicide.hr ~ dep.cons + grav.cons + suicide.past
    + past.prison + sczdys.cons + death + against.people
    + subst.cons + direct.miss + f.type.centre,
    family = "binomial", data = mhp.mod)

Deviance Residuals:
    Min        1Q       Median       3Q        Max
-2.08040   -0.46761   -0.22468   -0.08484   3.04981

Coefficients:
                      Estimate Std. Error z value Pr(>|z|)
(Intercept)           -6.8059     0.6774   -10.047  < 2e-16 ***
dep.cons               1.4106     0.2697     5.231  1.68e-07 ***
grav.cons              0.6727     0.1113     6.042  1.53e-09 ***
suicide.past           1.5755     0.2526     6.238  4.45e-10 ***
past.prison            0.6525     0.2593     2.516  0.011862 *
sczdys.cons            1.2968     0.7401     1.752  0.079726 .
death                  0.3508     0.2581     1.359  0.174216
against.people        -0.1648     0.2674    -0.617  0.537522
subst.cons            -0.2249     0.2822    -0.797  0.425496
direct.miss1           0.9873     0.2871     3.439  0.000584 ***❶
direct.missmissing     0.1801     0.3886     0.463  0.643013      ❷
f.type.centre2         0.4688     0.4236     1.107  0.268410
f.type.centre3         0.6242     0.3853     1.620  0.105193
---
Signif. codes: 0 '***' 0.001 '**' 0.01 '*' 0.05 '.' 0.1 ' ' 1

(Dispersion parameter for binomial family taken to be 1)

    Null deviance: 748.84 on 739 degrees of freedom
Residual deviance: 442.61 on 727 degrees of freedom
  (59 observations deleted due to missingness)
AIC: 468.61

Number of Fisher Scoring iterations: 6
```

Presenting a low level of self-directedness is a risk factor for being at
high risk for suicide attempt❶, and having missing data for the variable
"direct" is not a risk factor❷. This analysis is dually useful: (1) the 10%
missing observations for the variable "direct" are no longer discarded;
and (2) it can happen that having a missing value to a question is itself
a risk factor (e.g., no blood alcohol level available after a car accident can
be related to a car accident that was so severe that it was not possible to
collect the blood sample). However, this approach should be used with

caution because it may lead to biased coefficients, or at least to results that are difficult to interpret (Harrel 2001; Jones 1996): If subjects with missing values are among the most severe cases, as in the car accident example, the odds-ratio corresponding to the positive subjects will be underestimated (because precisely the most severe subjects are coded as "missing" and not as "positive").

To deal with the issue of missing values for the "direct" variable, it is also possible to impute the missing information. The simplest way to impute consists of assigning the median of the variable "direct" to all missing observations. The function impute() in the package "Hmisc" can be used for this purpose:

```
> library(Hmisc)
                    ❶
> mhp.mod$direct.median <- impute(mhp.mod$direct)
      ❷
> table(mhp.mod$direct.median, mhp.mod$direct, useNA = "ifany")

        0     1    <NA>
  0   567     0     85
  1     0   147      0
```

In ❶, the new variable direct.median is created; it is obtained from direct with NA values replaced by median(direct). This is verified in ❷ ("NA" in direct has been replaced by "0" in direct.median). This approach is robust and easy to use; however, it should only be used for preliminary or very exploratory analyses because it is potentially biased (Harrel 2001) and because more satisfactory alternatives exist, among which the function mice() in the "mice" library.

The function mice() iteratively and simultaneously imputes all missing values occurring in the dataset. It uses all the available information to replace missing values by the most probable ones. As explained in the preceding *"In a few words:"* paragraph, some noise is also introduced into the imputation process; in this way, two runs of mice() will generate two slightly different imputed datasets. By default, mice() creates five imputed datasets that can then be used in a multiple imputation technique:

```
> library(mice)
> impute.mhp <- mice(mhp.mod, seed = 1)

iter imp variable
                ❶
1    1   h.end.intervError in solve.default(t(xobs) %*% xobs) :
            system is computationally singular: reciprocal
            condition number = 2.65297e-22
```

```
                                                   ❷
> mhp.mod2 <- subset(mhp.mod, select = c(-centre, -type.centre2))
                              ❸
> impute.mhp <- mice(mhp.mod2, seed = 1)

iter imp variable
1    1    h.end.interv m.end.interv h.begin.interv m.begin.interv
          grav.cons n.child school prof.yn year.prison remand
          criminal duration work.prison discip past.prison
          n.prison d.prison
.... [TRUNCATED]
                          ❹
> mhp.mod.imp <- complete(impute.mhp)
#> describe(mhp.mod, xname = "mhp.mod")
#> describe(mhp.mod.imp, xname = "mhp.mod.imp")
   ❺
> table(mhp.mod.imp$remand, mhp.mod$remand, useNA = "ifany")

        1     2    3   <NA>
   1   219    0    0    0
   2    0   554    0    3
   3    0    0   23    0
   ❻
> table(mhp.mod.imp$prof.yn, mhp.mod$prof.yn, useNA = "ifany")

        0     1   <NA>
   0   230    0    2
   1    0   564    3
```

Unless caution is exercised, `mice()` is likely to lead to an error message as in ❶. In general, this is because `mice()` uses regression models, so that if there is any redundancy in the dataset, the estimation process comes to a halt due to a problem of multi-colinearity (see Section 5.1). In the present example, the variable "f.type.centre" is a factor obtained from a concatenation of the variables "centre" and "type.centre.2," and all three variables are present in the dataset. The last two variables should therefore be discarded and this is done in ❷. The function `mice()` is then called again. As explained above, by default, five datasets are imputed and some noise is introduced into the imputation process. This can be problematic because it means that the analyses are not totally reproducible. To solve this problem, the seed of the random number generator is set to 1❸. The function `complete()` can be used to extract an imputed dataset called "mhp.mod.imp"❹. Its compatibility with the non-imputed dataset "mhp.mod" is verified on two variables in ❺ and ❻.

It is now possible to use the imputed dataset "mhp.mod.imp" to carry out some analyses. However, it is better to perform multiple imputations using the five imputed datasets. Consider, for example, that a model explaining "being at high risk for suicide attempt" was not possible in the previous sections due to a too large number of missing values. With a single call on the function glm.mids(), five logistic regressions will be estimated and the results pooled appropriately so that we obtain the following final results:

```
> modimp <- glm.mids(suicide.hr ~ abuse + separation + discip
  + age + n.child + n.brother + school + prof.yn + remand
  + ns + ha + rd + f.type.centre, data = impute.mhp,
  family = "binomial")
> options(digits=2)
> summary(pool(modimp))
```

	❶				❷	❸	❹	❺	❻
	est	se	t	df	Pr(>|t|)	lo 95	hi 95	missing	fmi
(Intercept)	-3.422	0.7933	-4.314	319	2.1e-05	-4.983	-1.8616	NA	0.084
abuse	0.675	0.2111	3.197	478	1.5e-03	0.260	1.0899	7	0.055
separation	0.080	0.1998	0.398	719	6.9e-01	-0.313	0.4717	11	0.019
discip	0.197	0.2288	0.860	509	3.9e-01	-0.253	0.6460	6	0.050
age	-0.011	0.0094	-1.167	741	2.4e-01	-0.029	0.0075	2	0.015
n.child	0.032	0.0620	0.517	715	6.1e-01	-0.090	0.1538	26	0.020
n.brother	0.080	0.0269	2.962	544	3.2e-03	0.027	0.1323	0	0.045
school	-0.124	0.1069	-1.160	709	2.5e-01	-0.334	0.0859	5	0.021
prof.yn	-0.253	0.2082	-1.217	694	2.2e-01	-0.662	0.1555	5	0.024
remand	0.192	0.2080	0.924	442	3.6e-01	-0.217	0.6010	3	0.061
ns	0.207	0.1261	1.643	87	1.0e-01	-0.043	0.4578	104	0.200
ha	0.519	0.1332	3.898	37	4.0e-04	0.249	0.7893	108	0.320
rd	0.036	0.1400	0.255	40	8.0e-01	-0.247	0.3185	114	0.304
f.type.centre2	0.019	0.3472	0.055	726	9.6e-01	-0.662	0.7009	NA	0.018
f.type.centre3	0.387	0.3377	1.145	718	2.5e-01	-0.276	1.0499	NA	0.020

```
Warning message:
In all(object$df) : coercing argument of type 'double' to logical
```

The coefficients are in ❶, the p-values in ❷, and the confidence intervals in ❸ and ❹. These results are interesting because they integrate all the information contained in the original dataset and because the p-values and confidence intervals are not affected (or only slightly) by the fact that some imputed information has been artificially created.

The number of missing observations in the original dataset for each variable is in ❺. The "fraction of missing information"❻ reflects the level of uncertainty for each coefficient that is due to missing data (Horton and Lipsitz 2001).

6.7 Bootstrap

In a few words: When a logistic regression model is estimated, it is possible to obtain confidence intervals for the regression coefficients using the `confint()` function. Formally, these confidence intervals are obtained from the second partial derivatives of the log likelihood function (Hosmer and Lemeshow 1989). This approach is not very intuitive and, in addition, it is correct only in certain circumstances, as seen in Section 5.3. Logistic regression models have conditions for validity, and it is difficult to be completely confident with them; in many situations, there are even serious doubts (e.g., when there are numerous predictors and an outcome with few events).

There is, however, a more intuitive approach to confidence intervals (Efron and Tibshirani 1993). Working on a sample of size n obtained from a given population, it is possible to replicate the study in a new sample of size n, and to estimate a new series of coefficients. If such replications are performed 10,000 times (for example), then for each coefficient there will be 10,000 estimates. For a given coefficient, if the 2.5% lowest and largest estimates are discarded, then a 95% confidence interval is obtained. Of course, this approach is unrealistic: Nobody will replicate the same experiment 10,000 times to obtain a 95% confidence interval. However, these replications can be simulated. It is indeed possible to create a virtual population that could have provided the sample of size n. A possible method for this is simply to duplicate the original sample a large (in theory an infinite) number of times. From this virtual population, 10,000 samples of size n are randomly selected and a logistic regression model is estimated. From the empirical distribution of these 10,000 estimates, 95% confidence intervals can then be obtained as described above. This is what the bootstrap procedure does.

The bootstrap has two important advantages: (1) it does not rely on impenetrable mathematical considerations, and (2) it is a non-parametric approach, so that estimates of standard deviations and confidence intervals are derived with only minimal assumptions.*

The bootstrap also has some disadvantages (Young 1994); in particular, it is difficult to use with repeated measurements.†

The bootstrap is nevertheless a major breakthrough that gives more robust results and compels the statistics user to think properly about two essential questions: What is a p-value, what is a confidence interval?

In Practice: In Section 5.1 we ran a linear regression model to explain the "duration of interview" variable. When checking the conditions of validity of this model, we noticed graphically that the residuals were not, in fact, normal. This casts slight doubt concerning the validity of the p-values and confidence intervals of the regression coefficients that were obtained as follows:

```
> mod <- lm(dur.interv ~ scz.cons + dep.cons + subst.cons
  + grav.cons + char + trauma + age + f.type.centre,
  data = mhp.mod)
> summary(mod)
```

---

* The main assumption is that the observations are independent and identically distributed.
† Because the observations are no longer independent.

```
Call:
lm(formula = dur.interv ~ scz.cons + dep.cons + subst.cons
    + grav.cons + char + trauma + age + f.type.centre,
    data = mhp.mod)

Residuals:
    Min        1Q     Median        3Q       Max
 -38.578   -13.855    -1.769    10.922    64.405

Coefficients:
                  Estimate  Std. Error  t value  Pr(>|t|)
(Intercept)       42.48996     3.99998   10.623  < 2e-16    ***
scz.cons           3.06420❶    2.80997    1.090  0.275911❷
dep.cons           6.71269     1.64793    4.073  5.21e-05   ***
subst.cons         4.60037     1.79854    2.558  0.010760   *
grav.cons          1.06236     0.56548    1.879  0.060737   .
char               1.62547     0.93536    1.738  0.082723   .
trauma            -0.66805     1.66931   -0.400  0.689144
age                0.20788     0.06066    3.427  0.000649   ***
f.type.centre2     4.09540     2.53401    1.616  0.106546
f.type.centre3    -1.29681     2.44159   -0.531  0.595509
---
Signif. codes:  0 '***' 0.001 '**' 0.01 '*' 0.05 '.' 0.1 ' ' 1

Residual standard error: 18.02 on 645 degrees of freedom
  (144 observations deleted due to missingness)
Multiple R-squared: 0.1086,  Adjusted R-squared: 0.09619
F-statistic: 8.734 on 9 and 645 DF,  p-value: 1.919e-12

> confint(mod)
                      2.5%        97.5%
(Intercept)     34.63540172  50.3445232
scz.cons        -2.45359014   8.5819818❸
dep.cons         3.47673774   9.9486331
subst.cons       1.06866080   8.1320760
grav.cons       -0.04803948   2.1727540
char            -0.21125077   3.4621966
trauma          -3.94598090   2.6098792
age              0.08876109   0.3269916
f.type.centre2  -0.88050669   9.0713035
f.type.centre3  -6.09122397   3.4976070
```

The coefficient of the variable "schizophrenia" is in ❶; it is not significantly greater than 0 because the p-value❷ is 0.28. Its 95% confidence interval is in ❸.

It is now possible to estimate a bootstrapped confidence interval of this coefficient using the boot() function in the "boot" package:

```
> library(boot)
                        ❹        ❺
> lm.boot <- function(data, index) {
                        ❻
mhp.boot <- data[index, ]
mod <- lm(dur.interv ~ scz.cons + dep.cons + subst.cons
    + grav.cons + char + trauma + age + f.type.centre,
    data = mhp.boot)
        ❼
coefficients(mod)
}
        ❽
> set.seed(10)
                        ❾
> resboot <- boot(mhp.mod, lm.boot, 10000)
                        ❿
> hist(resboot$t[, 2], breaks = 40, main = "",
    xlab = "Regression coefficient of variable scz.cons")
> box()
```

The first and trickiest step of a bootstrap procedure programmed in R consists of the definition of the function that gives the parameter(s) of interest from a new dataset sampled from the population. This function has two arguments: one labels the dataset ❹ and the other the observations that are resampled (most often, observations correspond to the rows of the dataset ❺). The resampled dataset is obtained in ❻. Then the linear regression model is estimated and the coefficients are obtained in ❼. Because the bootstrap is based on the random resampling of a dataset, it is potentially a non-reproducible technique (two statisticians working on the same data with the same instructions can obtain slightly different results due to the random resampling). To deal with this issue, it is a good habit to systematically initialise the random number generator with the instruction set.seed()❽; the estimation process then becomes perfectly reproducible. The boot() function is finally called in ❾ with, successively, the name of the original dataset, the function to bootstrap (defined previously), and the number of replicates. This number should be greater than 1,000 when confidence intervals are calculated (Efron and Tibshirani 1993), so that 10,000 should be enough in most situations. The histogram of the 10,000 estimates of the coefficient of "scz.cons" (the second coefficient ❿ because there is the intercept first) is presented in Figure 6.6. This histogram can be used to obtain 95% confidence intervals when the 2.5% smallest and largest coefficients are discarded. Non-parametric p-values can be obtained in a similar manner. These 95% confidence interval and p-values are here:

```
        ❶                    ❷                ❸
> boot.ci(resboot, index = 2, type = "bca")
BOOTSTRAP CONFIDENCE INTERVAL CALCULATIONS
Based on 10000 bootstrap replicates
```

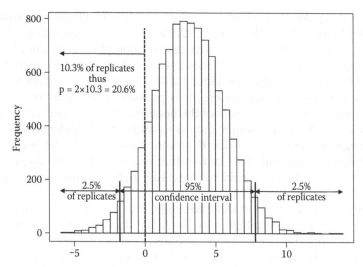

Regression coefficient of variable scz.cons

FIGURE 6.6

Histogram of the 10,000 estimates of a regression coefficient obtained from a bootstrap. The histogram approximates the coefficient distribution and can thus be used to estimate a 95% confidence interval or a p-value.

```
CALL:
boot.ci(boot.out = resboot, type = "bca", index = 2)
Intervals:
Level  BCa
95%    (-1.798❹, 7.881❺)
Calculations and Intervals on Original Scale
                    ❻
> p <- 2 * sum(resboot$t[, 2] < 0) / 10000
> p
[1] 0.2068❼
> p + 1.96 * sqrt(p * (1 - p) / 10000)
[1] 0.2147382❽
> p - 1.96 * sqrt(p * (1 - p) / 10000)
[1] 0.1988618❾
```

The function boot.ci()❶ is used to estimate a 95% confidence interval for the regression coefficient of the variable "scz.cons"❷. Several algorithms are available to derive this confidence interval; the option BCa (bias corrected accelerated) ❸ is sometimes considered preferable (Efron and Tibshirani 1993). The lower and upper bounds are in ❹ and ❺; they are close but not strictly equal to those obtained from the linear regression model (❸ and ❹ in the previous output). The one-sided p-value is obtained from the percentage of replicates that is smaller than 0 (because the original coefficient is positive) ❻ (Efron and Tibshirani 1993). The two-sided p-value is multiplied by 2 and is 0.206❼. Here again, it is close to, but distinct from, the parametric p-value obtained from the model (0.276❷ in the previous output). Because the bootstrap is based on an iterative constitution of random samples, it can give results that vary. The difference between ❼ (the bootstrapped p-value) and ❷ (the classic p-values provided by the linear regression) could thus be explained by random fluctuations. It is possible to test this hypothesis by calculating confidence intervals of the bootstrap p-value. This calculation is performed in ❽ and ❾ using standard formulae for confidence intervals of percentages. Clearly, the p-value ❷ obtained from the model is not within this confidence interval. This proves that one must be very cautious with parametric estimates of confidence intervals and p-values when conditions of validity are doubtful.

At this point, a remark should be made. The bootstrap considers explicitly that the population targeted by the MHP study is similar to an infinite number of replicates of the sample actually collected. In fact, this is not true. In the MHP study, there is a two-level random sampling procedure with unequal weights so that the sample does not set out to be a faithful representation of the overall population of French prisoners. We have seen this point in Section 4.1 and following; this is the reason why the package "survey" is of major interest. Although there has been some research on the extension of the bootstrap to multilevel sampling studies (Roberts and Fan 2004), the question is unfortunately not totally solved at the moment. To summarize, if the objective is to design a model that attempts to understand what happened *in a particular sample* and to take into account what could have been observed "by chance," then multilevel random sampling can be forgotten and the bootstrap can be used as presented above. Conversely, if the model is designed to make inferences about a population of interest (like the population of French prisoners here), then the survey package should be used and the bootstrap in this case is only for experienced users.

6.8 Random Effects and Multilevel Modelling

In a few words: Even if these words appear frequently in the literature related to questionnaire surveys, they are often considered esoteric concepts by researchers.

In fact, we have already done something similar to multilevel modelling without even noticing it. When we decided to take into account the two-level sampling design of the MHP study (e.g., using the "survey" package), we were, in a way, engaged in multilevel modelling. Indeed, prisoners from the same prison are likely to be somewhat similar, so that the variability of data collected in general will be less than the variability of data that would have been collected in a simple one-level sampling design (and the hypothesis of a one-level sampling is implicitly assumed in most statistical tests of hypotheses). It can be expected that this reduction in variability will lead, in practice, to an underestimation of standard deviations and to p-values that are artificially small. The functions svymean() or svyglm() in the "survey" package are used to correct these biases.

Random effects are another way to deal with the same issue. Here, a "prison" factor is included in all models to account for the similarity between prisoners from the same prison. But this "prison" factor will not have the same status as the other categorical predictors (e.g., the variable "profession"), and this is so for three reasons:

1. Because the variable "prison" has a large number of levels (20) and a predictor of this type can lead to inconsistent estimators, for example in non strictly linear models like logistic regression models (Davis 1985);

2. If the variable "prison" is included in the model, no more variables concerning the prison (e.g., date of construction, number of detainees, etc.) can be included in the regression model because the variable "prison" on its own monopolises all the information related to the prison where prisoners are held; and

3. If only a few prisoners have been sampled in a given prison, the corresponding "prison" effect will show considerable variance (Gelman and Hill 2008, p. 252).

For these three reasons, the "prison" variable will be considered a random-effect predictor, as opposed to the other explanatory variables like "profession," "age," etc. that are fixed-effect predictors. It is difficult to provide a strict definition of a random effect; as many as five different definitions have been found in the literature (Gelman and Hill 2008, p. 245). To put it simply, when a sample is divided into units, if these units have been randomly selected or if their number is greater than 10 (Snijders and Bosker 1999), the variable "unit" should be considered a random-effect predictor.

Now what is preferable: To use the "survey" package or to use a "prison" variable as a random-effect predictor? The answer to this question is not easy (Ghosh, Pahwa, and Rennie 2008; Lemeshow et al. 1998). The "survey" package is easy to use and deals simply with unequal sampling weights. Random effects are much more difficult to master, and the statistical routines necessary to estimate the corresponding models are still rather fragile (at least the present author is not totally confident about them at the moment). However, if one is interested in predictors that concern both the prisoners (age, history, etc.) and the prisons (size, rural or not, etc.), a model that allows for random-effect predictors is likely necessary.

In Practice: The function glmer() in the package "lme4" can be used to estimate linear, logistic, or Poisson regression models with fixed and random effects. If, for example, the objective is to explain the outcome variable "high risk of suicide attempt" with the fixed-effect predictors "history of child-hood abuse", "disciplinary procedure," "age," "type of prison," and the random-effect predictor "prison," then we can use the following syntax:

```
                        ❶                              ❷
> mod <- glmer(suicide.hr ~ abuse + discip + age + f.type.centre
       ❸                ❹
+ (1 | f.centre), data = mhp.mod, family = "binomial")
> summary(mod)
Generalized linear mixed model fit by the Laplace approximation
   Formula: suicide.hr ~ abuse + discip + age + f.type.centre
   + (1 | f.centre)
   Data: mhp.mod
   AIC   BIC   logLik deviance
   709   741.4 -347.5   695
Random effects:
   Groups    Name          Variance  Std.Dev.
   f.centre (Intercept) 0.53311❺  0.73014

Number of obs: 750, groups: f.centre, 20

Fixed effects:
                   Estimate  Std. Error  z value  Pr(>|z|)
(Intercept)       -1.846568   0.711670   -2.595   0.009467 **
abuse              0.747046   0.205137    3.642   0.000271 ***❻
discip             0.263741   0.231203    1.141   0.253981
age               -0.004804   0.008497   -0.565   0.571851
f.type.centre2    -0.085236   0.701560   -0.121   0.903300
f.type.centre3     0.421952   0.642140    0.657   0.511114
---
Signif. codes: 0 '***' 0.001 '**' 0.01 '*' 0.05 '.' 0.1 ' ' 1

Correlation of Fixed Effects:
             (Intr)   abuse    discip    age     f.ty.2
abuse        -0.144
discip       -0.229  -0.047
age          -0.540   0.031    0.272
f.typ.cntr2  -0.697   0.027   -0.012    0.002
f.typ.cntr3  -0.826   0.043    0.051    0.113   0.770
```

The instructions used to relate the outcome❶ to the fixed effect❷ are classic; this is not the case for the random effect "prison" ❸❹, the syntax of which is rather obscure. "1"❸ reflects the fact that the global level of suicide risk var-ies across prisons. Imagine that the investigators also require the relation-ship between age and suicide risk to vary across prisons: (1|f.centre)❸❹ should in this case be replaced by (age|f.centre). In ❹, f.centre results from the random selection of prisons participating to the study from the list all French prisons for men; this variable is therefore typically a random effect predictor.

Results concerning the fixed effects are presented in a traditional way ❻. A history of childhood abuse is the only predictor significantly associated with the outcome.

Results concerning the random effect are in ❺: There are no p-values and no coefficients, but instead we have the variance of the "prison" random effect. In a linear model, there would be a supplementary line with the residual variance. The proportion of the total variance (the total variance is the "prison" variance plus the residual variance) accounted for by the "prison" variance could be a convenient way to assess the magnitude of the "prison" effect. In a logistic regression model, if it is assumed that the dependent variable is based on a continuous unobserved variable (i.e., above a certain level, prisoners are considered to be "at high risk for suicide attempt"), then the residual variance is fixed and equal to $\pi^2/3 \approx 3.29$. The proportion of the total variance accounted for by the "prison" variance for the outcome variable "high risk of suicide attempt" is thus as follows:

❼
```
> icc <- 0.53311 / (0.53311 + pi^2/3)
> icc
[1] 0.1394489
```

where 0.53311 is obtained from ❺. ❼ is called the intraclass correlation coefficient; it corresponds to the residual correlation between "being at high risk of suicide attempt (y/n)" for two prisoners in the same prison. The intraclass correlation coefficient (icc) is important because it is needed to estimate the design effect d in a sampling process, which impacts the power of the study (see Section 4.8): $d = 1 + (n - 1) \times icc$, where n is the size of the sample recruited in each prison (Snijders and Bosker 1999). In practice, the design effect d is an estimate of the relative efficiency (i.e., relative sample size requirement) of simple random sampling compared to two-level random sampling (such as that used in the MHP study):

❽
```
> n <- mean(table(mhp.mod$centre))
> n
[1] 39.95
```
❾
```
> 1+icc*(n-1)
[1] 6.431534
```

The average number of prisoners sampled in each prison is estimated in ❽ and the design effect in ❾. Simple random sampling would have required about 6.5 fewer prisoners to obtain the same precision in the estimate of the prevalence of "being at high risk of suicide attempt." It is noticeable that design effects can vary substantially from one outcome to another: For example, concerning the diagnosis of alcohol abuse or dependence, d is equal to 1.37. In the power calculation for the MHP study, the design effect was postulated *a priori* to be 4.

Because there are two different approaches to managing multilevel designs (random effects and "survey" package), it is potentially interesting to take a look at results provided by the svyglm() function in the "survey" package:

```
                ❶
> mhp.mod$pw[mhp.mod$type.centre2 == 1] <- 1
> mhp.mod$pw[mhp.mod$type.centre2 == 2] <- 1
> mhp.mod$pw[mhp.mod$type.centre2 == 3] <- 1
> mhp.mod$strat.size[mhp.mod$type.centre2 == 1] <- 10000
> mhp.mod$strat.size[mhp.mod$type.centre2 == 2] <- 10000
> mhp.mod$strat.size[mhp.mod$type.centre2 == 3] <- 10000
> library(survey)
> mhp.survey <- svydesign(id = ~centre, strata = ~type.centre2,
    weights = ~pw, fpc = ~strat.size, data = mhp.mod)
> mod <- svyglm(suicide.hr ~ abuse + discip + age,
    design = mhp.survey, family = "binomial")
> summary(mod)

Call:
svyglm(suicide.hr ~ abuse + discip + age, design = mhp.survey,
    family = "binomial")

Survey design:
svydesign(id = ~centre, strata = ~type.centre2, weights = ~pw,
    fpc = ~strat.size, data = mhp.mod)

Coefficients:
            Estimate Std. Error t value Pr(>|t|)
(Intercept) -0.94371    0.48422  -1.949 0.071633 .
abuse        0.75366    0.18059   4.173 0.000938 ***❷
discip       0.25236    0.21828   1.156 0.266976
age         -0.01902    0.01229  -1.548 0.143825
---
Signif. codes: 0 '***' 0.001 '**' 0.01 '*' 0.05 '.' 0.1 ' ' 1

(Dispersion parameter for binomial family taken to be 0.9997305)

Number of Fisher Scoring iterations: 4
```

Because we did not consider the sampling weights in the glmer() call, uniform weights are used in ❶. The fixed effects are in ❷; they appear very close to those obtained previously in ❻ (it is not necessarily the case because the two approaches are basically different: the "survey" perspective uses a weighted average of the prison effect while the random effect approach fixes the level of this prison effect artificially (Graubard and Korn 1996)).

7

Principles for the Validation of a Composite Score

Many subjective concepts such as client satisfaction, problems with alcohol, or personality traits can be measured using lists of items that are aggregated secondarily into a global composite score. These lists of items are often called "scales." They are particularly interesting when complex constructs are to be measured. In this context, if an interviewee is asked a single question, the answer can be dubious. Conversely, if several items each tackle a simple, unambiguous aspect of the construct, then the summation of all the answers can provide a trustworthy measurement.

Most often, scales are designed and validated in specific studies, using sophisticated statistical procedures discussed in many books (Nunnally and Bernstein 1994). The objective here is more modest and it is definitely not to discuss the academic aspects of the validation of a psychometric tool, which raise many technical and theoretical questions. More simply, when a well-known scale is used in a questionnaire survey, certain basic properties concerning its validity should be verified again, in particular because the population studied is most often not strictly comparable to the population of the original validation study.

7.1 Item Analysis (1): Distribution

In a few words: Descriptive statistics should be presented for scale items just as for all other types of responses to a questionnaire (see Section 2.2). However, two points are of particular importance here: missing data and ceiling/floor effects.

The question of missing data is particularly sensitive because, if a score is obtained from the addition of several items, a single missing item is enough to make it impossible to estimate the total score.

Ceiling and floor effects are defined by a predominance of responses corresponding, respectively, to the lowest or the highest possible response levels. For example, if an item has a four-level response pattern coded "0," "1," "2," and "3," a floor effect will typically occur if 72% of subjects answer "0," 18% answer "1," 6% answer "2," and 4% answer "3." Symmetrically, a large proportion of "3" responses will correspond to a ceiling effect. Ceiling and floor effects are problems for at least two reasons:

1. The item distribution is far from normal; this can raise problems with statistical tools such as factor analysis, and

2. The item is poorly informative, as most subjects give the same response.

Under certain circumstances, an item with a floor or ceiling effect can nevertheless be interesting. This is the case, for example, for an item related to suicide ideations in a depressive scale: Few patients will give a positive answer, but this particular response is important from a clinical point of view and it enables discrimination of the most severe patients.

In Practice: As seen in Section 2.2, the functions `summary()` or `describe()` (package "prettyR") can be used to present the most salient elements of an item distribution. In this particular context, a graphical representation is, in general, more informative. Let us see, for example, how to represent the distribution of the 15 items of Cloninger's temperament scale (Cloninger 2000):

```
❶
> temp <- c("ns1", "ns2", "ns3", "ns4", "ns5", "ha1", "ha2",
  "ha3", "ha4", "ha5", "rd1", "rd2", "rd3", "rd4", "rd5")
                          ❷
> par(mfrow = c(3, 5))
         ❸              ❹                              ❺
> for(i in temp) barplot(table(mhp.mod[, i], useNA = "ifany"),
       ❻           ❼                    ❽
  main = i, space = 0.6, names.arg = c("1", "2", "3", "NA"),
           ❾                ❿
  ylim = c(0, 500), col = c("white", "white", "white", "black"))
```

The labels of all the temperament items are stored in a variable to facilitate the rest of the computation ❶. Fifteen diagrams must be placed on the same representation, and the layout is specified in ❷. By way of a loop, all item names are automatically called one after the other ❸. The functions `barplot(table())`❹ are used instead of `hist()` to ensure that all item levels are presented on each diagram, with the corresponding number of missing data ❺. The instruction `main = i` automatically gives the diagram the appropriate title. In ❼, the space between two bars is specified; and in ❽, the bars' labels are provided. It is important for the y-axis unit to be the same in all representations so that it is possible to compare the numbers from diagram to diagram, and this is the objective of ❾. Finally, it is specified in ❿ that the bar corresponding to missing data will be black.

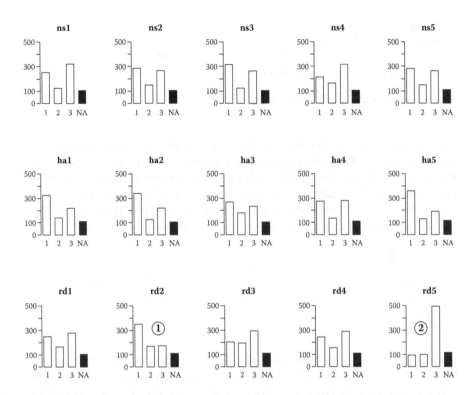

FIGURE 7.1
Distribution of the 15 items of a temperament scale ("ns" stands for novelty seeking, "ha" for harm avoidance, and "rd" for reward dependence). The number of missing data appears constant. Most distributions are "U"-shaped.

In Figure 7.1, it can be seen that most of items have a "U"-shaped distribution. That is, prisoners either present or do not present the trait; the intermediate level is infrequent. This results in a non-normal distribution of items. This is not necessarily a major problem here because it is temperament scores that are the parameters of interest and not the item values.

For each item, about 100 prisoners have missing data; we saw in Section 6.6 that these missing data involve the same group of prisoners. There is a floor effect in ① (rd2) and a ceiling effect in ② (rd5). A low level of rd2 corresponds to being "unresponsive or resistant to social pressure"; people in prison are indeed likely to present this feature. A high level of rd5 corresponds to "responsive to sentimental appeals or fond of saving nostalgia and memorabilia." Here again, this makes sense because it can be considered a side effect of imprisonment: From a technical point of view, one can say that item rd5 appears to be biased in the particular context of imprisonment.

Now let us take a look at the score for "novelty seeking," obtained from the summation of ns1, ns2, ..., ns5. First, the summation is performed and the number of prisoners with missing data for the score is estimated:

```
          ❶
> ns <- c("ns1", "ns2", "ns3", "ns4", "ns5")
                      ❷                        ❸
> mhp.mod$score.ns <- apply(mhp.mod[, ns], 1, sum)
      ❹
> sum(is.na(mhp.mod$score.ns))
[1] 110
```

The labels of the novelty-seeking items are stored in a vector in ❶. The function apply()❷ can be used to compute the summation of these items; the option in ❸ specifies that the columns are summed (the five items for a given subject), "2" would have summed up the rows. The number of missing values for the novelty-seeking score is computed in ❹: It is also around 100.

The histogram of the novelty seeking score is obtained as follows:

```
            ❶
> barplot(table(mhp.mod$score.ns), col = "white", main = "",
      ❷                                        ❸
  space = 0, xlab = "Novelty Seeking Score", ylim = c(0, 130))
      ❹
> box()
```

Here again, the barplot(table())❶ function is used instead of hist() to be sure that all novelty-seeking score values will have their own bar. In ❷, it is required that the bars be contiguous; ❸ deals with a bug introduced by box()❹; spontaneously, the box around the histogram may be put on top of the highest bar. In Figure 7.2, it is remarkable that some bars ❺ contain a number that is smaller than expected under a normality assumption. This is because most of the items have a distribution with a majority of "1" and "3" and a minority of "2"; the novelty seeking score is hence more often odd than even. If this score was to be used in a model designed for a normal random variable, the results should be viewed with caution.

7.2 Item Analysis (2): The Multi-Trait Multi-Method Approach to Confirm a Subscale Structure

In a few words: Most scales measuring polymorphous concepts such as quality of life or personality traits are, in fact, divided into subscales that assess more consistent and homogeneous facets of the phenomenon of interest. It can happen that these subscales are defined *a priori* by the author of the instrument. In all cases, their composition should be supported by data in the validation study (in general, using factor analysis; see Section 7.4). When such subscales are proposed for subjects from a population that differs from

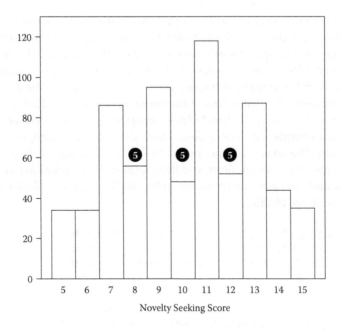

FIGURE 7.2
Histogram of the novelty seeking score. It appears that some bars ❺, regularly spaced contain numbers that are smaller than expected under a normality assumption. This surprising result comes from the "U" shape of the distribution of each item.

the initial validation population in terms of culture, age, and sometimes even language, an important question is: Is the initial subdivision of the scale still supported by data? For example, in a population of patients consulting a general practitioner, it can be shown that an instrument measuring quality of life is composed of five dimensions: physical functioning, bodily pain, vitality, social functioning, and mental health. It is not certain that the same structure will be observed in a population of depressed patients or in a population of patients with psoriasis (in particular, because physical functioning and bodily pain items will have different levels of variance and inter-item correlations). Sophisticated statistical methods such as confirmatory factor analysis can be used to deal with this issue; they are presented in Chapter 8. We focus here on a more simple and robust approach that is based on inter-item correlations and item-total score correlation. Indeed, if subscales are consistent, the inter-item correlations of items from the same dimension should be higher than the inter-item correlations of items from different dimensions. Likewise, the correlation between any one item and the total score of the subscale to which it belongs should be higher than the correlation between the same item and the total score of another subscale. This approach is in line with the development of the notions of convergent and divergent validity using the so-called "multi-trait multi-method" matrix (Campbell and Fiske 1959).

In Practice: The instrument developed by Cloninger to assess personality traits (Cloninger 2000) is divided into two parts. The first measures three dimensions of "temperament" that are thought to have a high level of genetic heritability: novelty seeking, harm avoidance, and reward dependence. Each of these three subscales consists of five items. The second part of the instrument measures four dimensions of "character" that are thought to emerge with time: self-directedness, cooperativeness, affective stability, and self-transcendence. These four subscales are also composed of five items. Using the mtmm() function in the "psy" package, it is possible to compare inter-item correlations and item-total score correlations in the temperament and character instruments in the MHP study sample. Concerning temperament, we obtain

```
> ns <- c("ns1", "ns2", "ns3", "ns4", "ns5")❶
> ha <- c("ha1", "ha2", "ha3", "ha4", "ha5")❶
> rd <- c("rd1", "rd2", "rd3", "rd4", "rd5")❶
                    ❷
> par(mfrow = c(1, 3))
      ❸                    ❹                    ❺
> mtmm(mhp.mod, list(ns, ha, rd), itemTot = TRUE,
    namesDim = c("Novelty Seeking", "Harm Avoidance",
    "Reward Dependence"))
```

	Item	ScaleI	1 Novelty Seeking	2 Harm Avoidance	3 Reward Dependence
1	ns1	1	❻0.223940808	❼0.101745419	❽0.04987198
2	ns2	1	0.423879492	0.097476733	0.04331521
3	ns3	1	0.400429302	0.128275597	0.08437935
4	ns4	1	0.308166274	-0.010411589	0.08115700
5	ns5	1	0.384362140	0.002740653	-0.02606461
16	ha1	2	0.092415979	0.474021279	0.14375621
17	ha2	2	0.014922239	0.516355321	0.18936090
18	ha3	2	0.054058538	0.456237973	0.06414960
19	ha4	2	0.151418198	0.370629528	0.05874885
20	ha5	2	0.037486824	0.416332518	0.13295379
31	rd1	3	-0.036335948	-0.115467997	0.33901499
32	rd2	3	0.049791095	0.267961771	0.28429907
33	rd3	3	-0.024379515	-0.028495778	0.40232556
34	rd4	3	0.220739437	0.247352486	0.21587665
35	rd5	3	0.002665781	0.163032514	0.30301885

The labels of the temperament items are grouped in three vectors corresponding to the three subscales ❶. The graphic window is divided into three parts (one per subscale) ❷. The mtmm() function is called in ❸; the main instruction is the list object that gathers the three vectors of labels ❹. The numerical results present the corrected item-total correlations (Nunnally

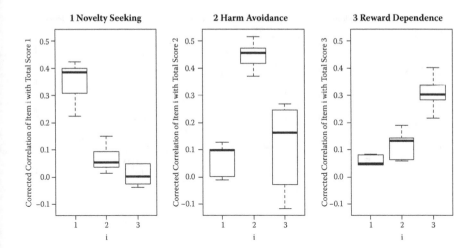

FIGURE 7.3

Corrected item-total correlations. In grey the item belongs to the subscale; in white the item belongs to another subscale. Grey boxes are clearly above white boxes: item-total correlations are higher when the item belongs to its predefined subscale. This is an argument for the validity of temperament subscales.

and Bernstein 1994, p. 304): For example, ❻ corresponds to the correlation of the first item of the novelty seeking subscale ("ns1") with the novelty seeking total score, from which ns1 is removed (to avoid the "overlap"). As expected, ❻ is greater than ❼ and ❽: The item-total correlation is higher with its own subscale. It is not easy to get a global perspective on these results; a boxplot obtained from the instruction in ❺ summarizes them efficiently (see, for example, Figure 7.3).

In Figure 7.3, the grey boxes (which correspond to the corrected correlation of items with their own subscale) are clearly above the white boxes (correlations of items with the other subscales); the breakdown defined *a priori* by Cloninger is thus supported by data collected in the MHP study. This is an argument for the validity of the temperament subscales. Because there are only five items per subscale, the boxplot in Figure 7.3 represents only five values for each. In this situation, "stripcharts" may be preferred (Figure 7.4). They can be obtained using the "stripChart" instruction:

```
> par(mfrow = c(1, 3))
> mtmm(mhp.mod, list(ns, ha, rd), itemTot = TRUE,
    stripChart = TRUE, namesDim = c("Novelty Seeking",
    "Harm Avoidance", "Reward Dependence"))
                   1 Novelty      2 Harm       3 Reward
     Item  ScaleI   Seeking      Avoidance    Dependence
1     ns1     1    0.223940808   0.101745419   0.04987198
.... [TRUNCATED]
```

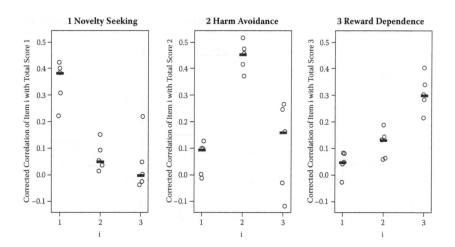

FIGURE 7.4

Stripcharts of the corrected item-total correlations. This representation could be preferred here to the boxplots proposed in Figure 7.3 because there are only five items per subscale (five values per boxplot).

The function mtmm() also systematically generates the representation of the inter-item correlations across subscales (Figure 7.5). The interpretation is similar to Figure 7.3:

What about the character scale? The function mtmm() now gives

```
> direct <- c("direct.1", "direct.2", "direct.3", "direct.4",
  "direct.5")
> coop <- c("coop.1", "coop.2", "coop.3", "coop.4", "coop.5")
> emostab <- c("emostab.1", "emostab.2", "emostab.3",
  "emostab.4", "emostab.5")
> transc <- c("transc.1", "transc.2", "transc.3", "transc.4",
  "transc.5")
> par(mfrow = c(1, 4))
> mtmm(mhp.mod, list(direct, coop, emostab, transc),
  itemTot = TRUE, namesDim = c("Self-directedness",
  "Cooperativeness", "Affective Stability",
  "Self-transcendence"))
.... [TRUNCATED]
```

The self-transcendence subscale seems to present a certain level of consistency (i.e., items correlate highly with the total score), but this is definitely not the case for the other subscales (see Figure 7.6). The original subdivision proposed a priori by Cloninger is thus not supported by the data here. This is not necessarily a drawback because Cloninger (2000) has (Cloninger 2000) suggested that the four character dimensions should be pooled to determine the severity of a personality disorder. The four subscales of character could therefore constitute a homogeneous family—in statistical terminology, this would correspond to a property of unidimensionality.

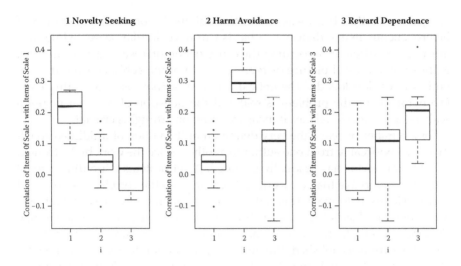

FIGURE 7.5
Distribution of inter-item correlations for the temperament subscales. The different items correlate better with other items from the same subscale (grey boxes are above white boxes). These results can be related to those presented in Figure 7.3.

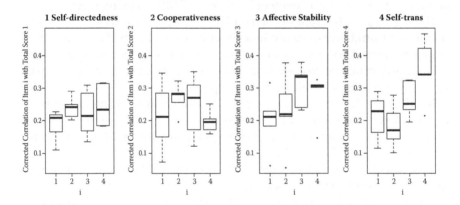

FIGURE 7.6
Corrected item-total correlations for the character subscales. Only self-transcendence seems to emerge from the pool of items. The original subdivision proposed by Cloninger is not supported by the data here.

7.3 Assessing the Unidimensionality of a Set of Items

In a few words: It is often pointed out by specialists that a scale should measure only one characteristic at a time: "[...] suppose a test of liberalism consists of two general sorts of items, one concerned with economic and the other with religious issues. Two individuals could thus arrive at the same numerical score by quite different routes." (McNemar 1946). This author suggests here that if a measurement is obtained from the aggregation of heterogeneous items, then two subjects can have similar scores while they are actually different concerning the construct of interest: The measurement is thus not a faithful picture or representation of the subjects. Even if economists have developed a theoretical framework that deals with this issue (Thurston 2006), the problem is real and the question of whether or not a scale is unidimensional (i.e., it measures only one thing) is crucial.

For a number of years, the very definition of the notion of unidimensionality varied noticeably from one author to another. Since the work by McDonald (1981), a consensus has gradually emerged: A set of items is unidimensional if there is a variable that, on its own, "explains" all the correlations observed between items. In formal terms, if this variable is maintained at a constant level in a sample of subjects, then items in this sample are independent of one another. This indeed reflects the existence of a "single core," common to all the items; once this core is established, the items behave randomly and independently of each other. Although this definition seems appropriate from a theoretical point of view, it is not sufficient in practice: The proportion of the test variance explained by the common factor must be large (Cronbach 1951). Indeed, if an instrument measures a single concept but that concept is swamped in a large amount of noise, the measurements will also be difficult to interpret.

The eigenvalue diagram or "scree plot" is an essential tool for the determination of unidimensionality in a set of items (Cattel 1966) because it can help determine if one, two, or more common factors are necessary to explain a large proportion of test variance. Let us see, from a set of geometrical representations, how results can be interpreted.

To take an extremely simple example, consider an instrument composed of three items administered to 27 subjects. As each subject undergoes three evaluations, this data can be represented geometrically by 27 dots in a three-dimensional space (Figure 7.7).

A principal components analysis (PCA) of these data consists of looking for the three directions in which the 27 subjects spread (Figure 7.8).

The 27 dots spread more or less along the principal components. Numerically, the square root of the so-called "eigenvalue" λ_1 associated with the first principal component indicates how closely the subjects are dispersed along this component; λ_1 also corresponds to a good approximation of the proportion of variance of the three variables x, y, and z explained by the most prominent common factor (Jackson 1991). The same is true for λ_2 and λ_3 in relation to the second and third principal components, which can be associated with the other principal components (all statistically independent). If λ_1 is much higher than λ_2 and λ_3, then the dots in Figure 7.7 will be situated more or less along a straight line. This means that they form a unidimensional whole.

A scree plot diagram is the simple graphic representation of the eigenvalues λ_1, λ_2, and λ_3 (Figure 7.9). When x, y, and z form a unidimensional set, λ_1 is considerably greater than λ_2 and λ_3 and this corresponds to Figure 7.10.

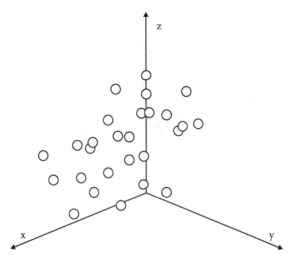

FIGURE 7.7
Geometrical representation of three items (x, y, and z) measured on 27 subjects.

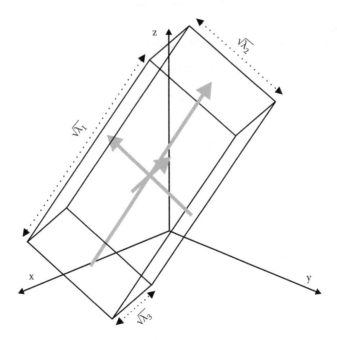

FIGURE 7.8
The three principal components of x, y, and z correspond to the directions in which the 27 subjects spread. The eigenvalue λ_1 associated with the first principal component indicates how closely the subjects are dispersed along this component (likewise for λ_2 and λ_3); λ_1 also corresponds to the proportion of variance explained by the most prominent common factor.

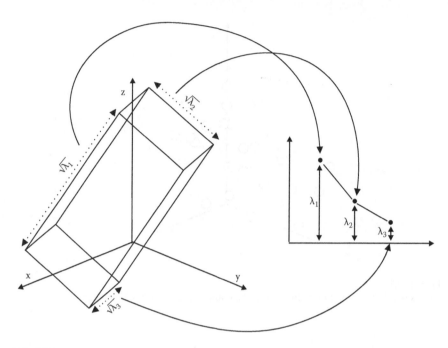

FIGURE 7.9
Construction of a scree plot that presents the three eigenvalues λ_1, λ_2, and λ_3.

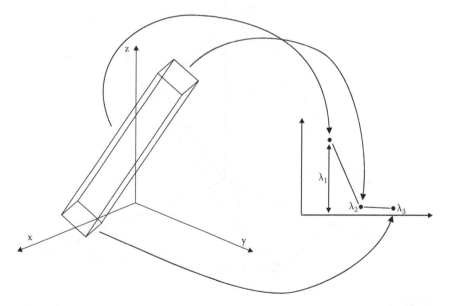

FIGURE 7.10
When λ_1 is much larger than λ_2 and λ_3, the subjects are approximately positioned along a straight line, and a common factor on its own explains a large proportion of variance: x, y, and z form a unidimensional set of variables.

In Practice: The function screeplot() in the "psy" package can be used to draw a scree plot. If we consider the items in the character scale:

❶
```
> scree.plot(mhp.mod[, charac], simu = 20,
    title = "Scree Plot of Character Variables")
```

In Figure 7.11, the first eigenvalue ❷ is well above the 19 others. This is in favor of the unidimensionality of the scale. But how can it be decided whether the second eigenvalue ❸ is negligible? More generally, when reading a scree plot, how can the number of dimensions that have a significant role in explaining the variance of the twenty items be determined? This important question is regularly debated in the literature, and there are today the beginnings of a consensus on the value of simulations (Lance, Butts, and Michels 2006). This explains the use of simu = 20 in ❶. Twenty datasets similar in size to mhp.mod[, charac] are randomly generated and the 20 corresponding scree plots are added to Figure 7.11 (❹). These scree plots give an idea of "what could be expected by chance" so that it appears that two and perhaps three eigenvalues seem to be significant. This approach is called "parallel analysis" (Horn 1965). Most often, the "Keyser rule" is used instead of parallel analysis. This rule suggests that eigenvalues larger than 1❺ should be retained, and here there are five. The rationale in this approach is that the summation of all the eigenvalues is equal to the number of items so that, on average, an eigenvalue is equal to 1. Eigenvalues larger than 1 are thus assumed to be "above the average." Simulations have shown that this approach overestimates the number of significant dimensions (Lance, Butts, and Michels 2006).

To summarise: Concerning the character scale, it appears that a common factor on its own explains a substantial amount of the item variance and that no other factor appears of comparable importance, even if one or perhaps two other factors have an amount of variance above what could be expected by chance (and it might be interesting to determine what these other factors correspond to using factor analysis; see Section 7.4). The scale is thus fairly homogeneous, if not strictly unidimensional. It is thus admissible to sum all 20 items (and this is implicitly what Cloninger has suggested (2000)).

Now what can be said of the temperament scale?

```
> scree.plot(mhp.mod[, temp], simu = 20,
    title = "Scree Plot of Temperament Variables")
```

In Figure 7.12, the first eigenvalue is not obviously and uniformly above the 14 others. Instead, it appears that the first three eigenvalues are above the 20 simulated scree plots. Here, the interpretation is that the temperament scale is not unidimensional, but rather tridimensional, so that from a

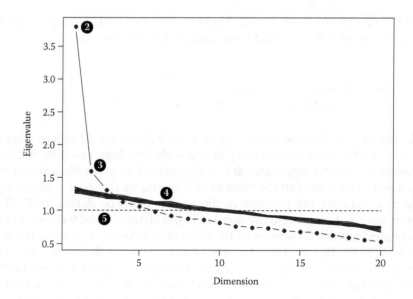

FIGURE 7.11
Scree plot of the items in the character scale. The first eigenvalue is well above the 19 others; the instrument is approximately unidimensional. Two and perhaps three eigenvalues are above "what could be expected by chance"❹.

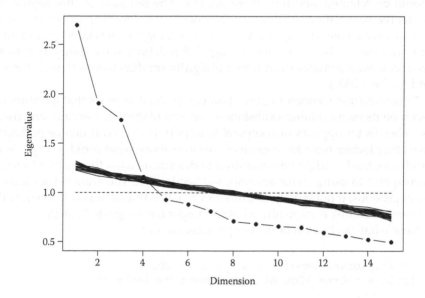

FIGURE 7.12
Scree plot of the items in the temperament scale. The first eigenvalue is not obviously and uniformly above the 14 others. It appears, instead, that three eigenvalues are clearly above "what could be expected by chance." The temperament scale is thus not unidimensional; it should be divided into three subscales.

theoretical point of view, it is not recommended to gather the items into a single total score. The scale should instead be divided into three subscales, and the question becomes: Which items should be grouped together?

7.4 Factor Analysis to Explore the Structure of a Set of Items

In a few words: Once it appears that a scale is structured in two, three, or more dimensions, it is useful, in general, to determine which items belong to which dimension. This is the objective of factor analysis. From a formal point of view, factor analysis is a linear model that explains the responses to items by a few latent, unobserved variables. Consider, for example, that five psychological tests are administered to 112 individuals; the results of these tests are associated with the following variables (Smith and Stanley 1983): "picture" (a picture-completion test), "blocks" (block design), "maze," "reading" (reading comprehension), and "vocab" (vocabulary). A two-factor factor analysis model applied to these data is defined by

$$\text{picture}_i = a_1 + b_1\,F_i + c_1\,G_i + e^1_{\ i}$$

$$\text{blocks}_i = a_2 + b_2\,F_i + c_2\,G_i + e^2_{\ i}$$

$$\text{maze}_i = a_3 + b_3\,F_i + c_3\,G_i + e^3_{\ i} \qquad (7.1)$$

$$\text{reading}_i = a_4 + b_4\,F_i + c_4\,G_i + e^4_{\ i}$$

$$\text{vocab}_i = a_5 + b_5\,F_i + c_5\,G_i + e^5_{\ i},$$

where i runs from 1 to 112, F and G designate two unknown independent characteristics common to the five tests (F and G are the common factors), and e corresponds to the measurement error plus the part of response specific to a given test. The coefficients a_i, b_i, and c_i are traditionally called "loadings." Now, if the loadings b_1, b_2, b_3 are high while b_4, b_5 are low (i.e., close to 0) and, conversely, if c_1, c_2, c_3 are high while c_4, c_5 are low, then two important conclusions can be inferred.

First, it can be said that "picture," "blocks," and "maze" are mainly explained by the same common predictor, while "reading" and "vocab" are mainly explained by another predictor. Second, it is now possible to imagine a meaningful label for the unmeasured F and G. Indeed, a high score in "picture," "blocks," and "maze" is likely to correspond to a high level in something like "performance intelligence" (F), while a high score in "reading" and "vocab" could correspond to a high level in "verbal intelligence" (G).

It may seem strange to design a model in which the predictors are unknown. This is, in fact, a long tradition in mathematics, and it is indeed possible to estimate the a_i's, b_i's, and c_i's from the 112 scores in the five tests that make the equations in (7.1) "as true as possible" (e.g., using the maximum likelihood principle or a least-squares algorithm). Unfortunately, once an optimal solution for the loadings (a_i, b_i, c_i)'s has been found, it can be proven that there are an infinite number of solutions of the same quality. Each solution can be obtained from another by means of a mathematical operator formally equivalent to a rotation. This problem of underdetermination in factor analysis is regularly presented as a major drawback. This is somewhat disproportionate, as even if there are an infinite number of solutions, they generally all lead to the same pattern of items per dimension and to the same labelling for the common factors. Two particular rotations are traditionally favoured: (1) varimax, in which the common factors remain independent; and (2) promax, which allows them to be correlated. These two rotations help in the interpretation of results because they lead to coefficient estimates that are as close as possible either to 0 or to 1.

Finally, it can be noted that, in theory, factor analysis is valid on the condition that the dependent variables have a normal distribution. However, this condition of validity should not be considered too strictly because most of the time no statistical test of hypotheses will be performed.

In Practice: In the previous section, the scree plot showed that the temperament scale consists of three dimensions. What are the three corresponding subscales? It is possible to answer this question using factor analysis and the function factanal():

```
                              ❶                        ❷
> print(factanal(na.omit(mhp.mod[, temp]), factors = 3),
            ❸
    cutoff = 0)


Call:
factanal(x = na.omit(mhp.mod[, temp]), factors = 3)

Uniquenesses:❹
  ns1   ns2   ns3   ns4   ns5   ha1   ha2   ha3   ha4   ha5   rd1   rd2
0.928 0.635 0.573 0.845 0.773 0.633 0.564 0.673 0.776 0.721 0.574 0.774
  rd3   rd4   rd5
0.576 0.805 0.840


Loadings:
        Factor1    Factor2    Factor3
ns1     0.065      0.259❺     0.023
ns2     0.062      0.601❺    -0.025
ns3     0.129      0.639❺    -0.038
```

```
ns4    -0.040      0.386❺    0.065
ns5    -0.027      0.468❺   -0.084
ha1     0.602❻     0.063     0.035
ha2     0.644❻    -0.045     0.140
ha3     0.570❻     0.004    -0.052
ha4     0.452❻     0.127    -0.054
ha5     0.526❻     0.037     0.030
rd1    -0.151     -0.033     0.634❼
rd2     0.303      0.112     0.348❼
rd3    -0.057     -0.017     0.649❼
rd4     0.281❼     0.275     0.201
rd5     0.165     -0.012     0.364❼

                  Factor1   Factor2   Factor3
SS loadings        1.834     1.317     1.159
Proportion Var     0.122❽    0.088     0.077
Cumulative Var     0.122     0.210     0.287

Test of the hypothesis that 3 factors are sufficient.❾
The chi square statistic is 163.07 on 63 degrees of freedom.
The p-value is 8.25e-11
```

The function na.omit()❶ is used to discard all observations with missing data. It is a good habit to verify, as we did in Section 7.1, that the number of subjects available for the analysis is sufficiently large. If not, an imputation of missing items can be discussed (see Section 6.6). The number of common factors is specified in ❷. By default, small loadings are not displayed in the output. This is supposed to facilitate the interpretation of results. From a scientific point of view, it is likely preferable to have all the values, and the instruction cutoff = 0 is used for this purpose. In ❹, for each item, we have the "uniquenesses," which are defined as the proportion of variance not explained by the three common factors. Then the loadings estimated after a varimax rotation are given. Loadings correspond to the common factor coefficients; they lie between 1 and −1, and can also be interpreted as the Pearson correlations between the items and the common factors.

As a first step, it is useful to look for the common factor that has the strongest association with each item.* All novelty seeking items have their largest loading on factor 2 ❺; all harm avoidance items have their largest loading on factor 1 ❻; and all reward dependence items except rd4 have their largest loadings on factor 3 ❼. The situation is thus clear: factor 1 can be labeled "harm avoidance," factor 2 "novelty seeking," and factor 3 "reward dependence."

---

* A strong association that can be positive or negative; in other words, a loading close either to 1 or to −1.

Second, it may be useful to look for the items that are a distinctive characteristic of a given common factor: For novelty seeking, this is typically the case of ns2, ns3, and ns5, which have high loadings on factor 2 and very small loadings on the other factors. In contrast, concerning reward dependence, rd2, rd4, and to a lesser extent rd5 have small loadings on their related factor and, for rd2 and rd4, loadings of a similar magnitude on other factors. To interpret the rather weak structure of factor 3, we need to return to the labelling of the items. We already noticed in Section 7.1 that items rd2 "unresponsive or resistant to social pressure" versus "depends on emotional support from others, or yields easily to social pressure" and item rd5 "unresponsive to sentimental appeals, and not nostalgic about memorabilia" versus "responsive to sentimental appeals or fond of saving nostalgia and memorabilia" could have a particular meaning in prison. Concerning item rd4 "insensitive to rejection or criticism" versus "sensitive to rejection or criticism," it is possible that the notion of "sensitivity" could have been interpreted by interviewers as a paranoid trait, which is not rare in prison.

To summarise, factor analysis of the temperament scale qualitatively confirms the original structure in three subscales as proposed by Cloninger (2000). The reward dependence factor, however, seems less robust than the others.

In addition to the loadings, the proportion of variance explained by each factor is given in ❽. A goodness-of-fit test is provided in ❾. This test is not really meaningful here. First, factor analysis should be considered an unpretentious exploratory method that can help determine which items belong to which factor and nothing more. Second, factor analysis requires items that are normally distributed, and this is obviously not the case here, as each question concerning the temperament scale can be answered by "high," "average," or "low." Although normality is not, in general, an absolute prerequisite for an exploratory tool (even if some authors regularly raise problems related to factor analysis carried out on binary or ordered items (Nunnally and Bernstein 1994, p. 570)), it does become essential when statistical tests of hypothesis are performed.

A question can be raised at this point: Why was the orthogonal rotation varimax chosen here?

1. Because from a theoretical point of view the temperament dimensions are thought to be independent.

2. Because results obtained with an oblique rotation like promax are less easy to interpret. Loadings are not Pearson correlations between items and factors, the proportion of variance explained by each factor cannot be interpreted simply, and so forth. (Nunnally and Bernstein 1994, p. 534; Tacq 1997, p. 282).

Nevertheless, if an oblique rotation is desired, the function `promax()` can be used with the function `print.psych()` in the "psych" package to obtain the correlation between factors:

```
> res <- factanal(na.omit(mhp.mod[, temp]), factors = 3)
> promax.load <- promax(loadings(res))
> library(psych)
> print.psych(promax.load, cut = 0)
```

Call: NULL

	item	Factor1	Factor2	Factor3	h2	u2
ns1	1	0.02	0.26❶	0.02	0.07	0.93
ns2	2	-0.05	0.62❶	-0.03	0.38	0.62
ns3	3	0.01	0.65❶	-0.05	0.43	0.57
ns4	4	-0.11	0.40❶	0.08	0.18	0.82
ns5	5	-0.12	0.49❶	-0.07	0.26	0.74
ha1	6	0.61❷	0.02	-0.06	0.37	0.63
ha2	7	0.67❷	-0.10	0.04	0.46	0.54
ha3	8	0.59❷	-0.03	-0.14	0.37	0.63
ha4	9	0.44❷	0.10	-0.12	0.22	0.78
ha5	10	0.54❷	0.00	-0.05	0.29	0.71
rd1	11	-0.15	-0.05	0.66❸	0.46	0.54
rd2	12	0.29	0.08	0.30❸	0.18	0.82
rd3	13	-0.06	-0.04	0.66❸	0.44	0.56
rd4	14	0.24	0.25❸	0.16	0.15	0.85
rd5	15	0.17	-0.04	0.34❸	0.15	0.85

	Factor1	Factor2	Factor3
SS loadings	1.87	1.37	1.16
Proportion Var	0.12	0.09	0.08
Cumulative Var	0.12	0.22	0.29

With factor correlations❹ of

	Factor1	Factor2	Factor3
Factor1	1.00	0.26	0.17
Factor2	0.26	1.00	0.07
Factor3	0.17	0.07	1.00

The factor pattern in ❶, ❷, and ❸ is similar to the one obtained with the varimax rotation. The factors are no longer independent; however, the correlations between them are rather weak ❹.

Now, what can be said about the structure of the character scale? A factor analysis with a varimax rotation gives

```
> print(factanal(na.omit(mhp.mod[, charac]), factors = 4),
    cutoff = 0)
```

Call:
```
factanal(x = na.omit(mhp.mod[, charac]), factors = 4)
```

```
Uniquenesses:
direct.1   direct.2    direct.3    direct.4    direct.5     coop.1
  0.858      0.945       0.798       0.734       0.781       0.851
 coop.2     coop.3      coop.4      coop.5     emostab.1   emostab.2
  0.646      0.814       0.704       0.762       0.663       0.680
emostab.3 emostab.4   emostab.5   transc.1    transc.2    transc.3
  0.835      0.661       0.761       0.633       0.548       0.772
transc.4   transc.5
  0.763      0.885
```

Loadings:

	Factor1	Factor2	Factor3	Factor4
direct.1	0.156	0.051	0.081	0.330[1]
direct.2	0.026	0.128	0.077	0.178[1]
direct.3	0.115	0.392[1]	0.180	-0.053
direct.4	0.013	0.431[1]	0.119	0.257
direct.5	0.363[1]	0.135	0.097	0.245
coop.1	0.276[2]	0.076	-0.003	0.258
coop.2	0.023	-0.089	0.006	0.588[2]
coop.3	0.302[2]	0.187	0.182	0.165
coop.4	0.511[2]	0.085	0.079	0.146
coop.5	0.300	-0.014	0.039	0.383[2]
emostab.1	0.127	0.545[3]	0.154	0.007
emostab.2	0.482[3]	0.205	0.123	0.175
emostab.3	0.308[3]	0.258	0.047	0.032
emostab.4	0.515[3]	-0.136	0.229	0.050
emostab.5	0.452[3]	0.124	0.138	0.025
transc.1	0.129	0.342	0.474[4]	0.092
transc.2	0.125	0.177	0.600[4]	0.211
transc.3	0.263	0.123	0.379[4]	0.013
transc.4	0.168	0.299	0.341[4]	0.050
transc.5	0.074	0.098	0.202	0.242[4]

	Factor1	Factor2	Factor3	Factor4
SS loadings	1.638	1.135	1.108	1.026
Proportion Var	0.082	0.057	0.055	0.051
Cumulative Var	0.082	0.139	0.194	0.245

```
Test of the hypothesis that 4 factors are sufficient.
The chi square statistic is 161.13 on 116 degrees of freedom.
The p-value is 0.00357
```

This situation is typical of a scale with a weak structure, or even with no structure at all. Most of the loadings are low, many items have loadings of a comparable magnitude on several factors, and the anticipated structure is not confirmed (except perhaps for the dimensions "emotional stability" and "self-transcendence"). In such circumstances, it is possible to estimate a factor analysis model with three of the five factors instead of four, or to use a promax rotation instead of varimax. A wiser attitude would indeed be to realise that the instrument has no clear-cut structure and hence that even if the four character subscales have theoretical interest, their relevance is not supported by the data.

7.5 Measurement Error (1): Internal Consistency and the Cronbach Alpha

In a few words: In most scientific fields, researchers work continually to improve their instruments so that measurement errors are gradually reduced. Does this notion apply to questionnaire studies? When a subject answers a question in a survey, does it make sense to consider something that could correspond to measurement error? Yes and no. On the one hand, if the investigator focuses only on what the subject says, there is by definition no measurement error (if technical problems, with data capture for example, are excluded). On the other hand, if the investigator is interested in something more ambitious like the "truth," the "subject's reality," then there can indeed be measurement error. Sometimes the "truth" does exist (e.g., "How many times did you smoke some cannabis in the last month?"), but it may be that the subject does not understand the question, does not remember very well, or is consciously or unconsciously embarrassed by the answer and therefore gives an erroneous response. Sometimes there is even doubt as to whether a "right answer" actually exists (e.g., "Are you satisfied with your life? Answer by circling a number between 0 and 10."). Here, the notion of measurement error becomes a real philosophical problem.

If measurement error associated with responses to a questionnaire is considered a relevant notion, how can it be assessed? If a gold standard is potentially available, an ancillary study can attempt to explore how far it correlates with the subject's answers. For instance, if an investigator has some doubts about the reliability of responses to the question, "What is your weight (in kilograms)?", it may be useful (but perhaps discourteous) to select about 10% or 5% of the subjects and to ask them to step on the scales once the interview terminates. Even if such validation studies generate interest and are regularly published (Strauss 1999), most survey questions cannot be related to a gold standard.

An alternative could be to repeat the measurement several times and to estimate to what extent the result is reproducible. Unfortunately, for a questionnaire, reproducibility is likely uninterpretable: If the answer is always the same, it is perhaps because the subject is making no effort and just repeating what he (she) said earlier, and not because the answer actually reflects what he (she) believes or thinks. The problem of measurement error in a questionnaire item thus appears insoluble. Surprisingly, this is not altogether true. Psychometricians have found how to get around this difficulty when the measurement is obtained from a composite scale. If the scale is divided randomly into two parts, the correlation of the two corresponding subscales corrected for length will give an idea of measurement reproducibility (there is a need to correct for length because a long test is, in general, more reproducible than a short one). Of course, these split-half coefficients will vary slightly according to the random division of the scale, so that it might be useful to estimate the mean of all possible split-half coefficients. This last parameter is called Cronbach's α coefficient (Nunnally and Bernstein 1994, p. 254).

In Practice: The character scale proposed by Cloninger (2000) provides a total score that can be used to determine the level of personality disorder in a given subject. The α coefficient of this scale can be estimated with the `cronbach()` function in the package "psy":

❶
```
> cronbach(mhp.mod[, charac])
$sample.size
[1] 671

$number.of.items
[1] 20

$alpha
[1] 0.767102❷
```

The vector charac❶ contains the names of the 20 items in the character scale. The result is in ❷: 0.767. It is equal to the mean of all the split-half correlations; it can also be interpreted as "what could be expected" if Cloninger's (2000) character scale is correlated with another conceptually equivalent character scale, but with a different content. Is 0.767 an acceptable value? As usual, it is impossible to answer this question seriously. Nevertheless, for a questionnaire used in a survey, a coefficient as high as 0.7 is acceptable (Nunnally and Bernstein 1994, p. 265).

Concerning the novelty seeking subscale in the temperament scale, we have

```
> ns <- c("ns1", "ns2", "ns3", "ns4", "ns5")
> cronbach(mhp.mod[, ns])
$sample.size
[1] 689

$number.of.items
[1] 5

$alpha
[1] 0.5929088❶
```

The result is smaller ❶: 0.59. This was expected because the novelty seeking subscale contains only 5 items, while the character scale has 20 items (a high α value can be obtained from a long scale or from a scale with large inter-item correlations). The measurement is less reliable, but this is not so bad, considering the small size of the scale.

7.6 Measurement Error (2): Inter-Rater Reliability

In a few words: Questionnaire surveys can be administered via the Internet or by post, but it often happens that it is an interviewer who asks the questions either by phone or during a face-to-face interview. The presence of an interviewer increases the budget of the study; however, it is sometimes necessary, for instance to ensure that subjects do not skip any item or that they have correctly understood the questions. It might happen that the interviewer has to interpret what he/she sees and what he hears. In all these situations, the interviewer is part of the measurement process and can thus be a source of measurement error.

A simple way to deal with this issue is at some time in the analysis to introduce an "interviewer effect" in the regression models. If this effect is significant, it is then useful to see which interviewers are outliers and to understand why they are outliers (perhaps they intervened in a particular area, they were older, etc.). This implies that some minimal information should be collected concerning the interviewers, which is not necessarily frequent.

Another approach consists of an estimation of inter-interviewer (or inter-rater) agreement; this is especially necessary when a large margin of interpretation is left to the interviewers. This can be done, for example, using videotaped interviews that are then rated by a second group of interviewers. If discrepancies are too large, the interviewer training sessions that are systematically delivered before the beginning of the study should be reinforced and adapted in consequence.

Inter-rater agreement concerning binary and, more generally, categorical variables (ordered or not) is traditionally estimated using a "kappa" coefficient. Consider first a situation where two interviewers must answer a "yes/no" question (Table 7.1).

The two interviewers gave the same answer in a proportion of $(a + b)/n$ subjects; this value is, by definition, the "concordance" of interviewer 1 and interviewer 2. This parameter is useful but it has an important drawback: It can be substantially greater than 0% simply by chance. For instance, if both interviewers give their answers by tossing "heads or tails?", they will have a concordance around 50%. Thus, it can be suggested that a concordance that takes into account agreement occurring by chance should be used, and this led to the definition of Cohen's kappa (Cohen 1960):

$$kappa = \frac{\text{Observed agreement} - \text{Chance agreement}}{1 - \text{Chance agreement}}$$

TABLE 7.1

Results of an inter-interviewer agreement study concerning a (yes/no) question. For a subjects, the interviewers both answered "yes"; for d subjects, neither did; and for c and b subjects, one interviewer answered "yes" but not the other. The total number of subjects is n = a + b + c + d.

| | | Interviewer 1 | |
		yes	no
Interviewer 2	yes	a	b
	no	c	d
			n

When there is perfect agreement, kappa = 1; and when there is agreement compatible with chance, kappa = 0. Of course, the problem now consists of the definition of "agreement occurring by chance." Cohen proposed that it should correspond to the situation where the two interviewers are independent in their answers; that is, the first interviewer answers "yes" with a probability of (a + c)/n and the second answers with a probability of (a + b)/n. This proposition has been criticized (Uebersax 1987), in particular because it gives paradoxical results when one interviewer has a large majority of "yes" and the other a large majority of "no" so that Cohen's kappa should not be estimated in this situation (Zwick 1988).

Cohen's kappa has been generalized to all types of categorical variable (binary or not, ordered or not (Graham and Jackson 1993)) and to the situation where there are more than two raters (Conger 1980). For a continuous variable, a different but compatible approach should be considered: that is, the intra-class correlation coefficient (Shrout and Fleiss 1979).

In Practice: In the MHP study, the role of interviewers was crucial; the Cloninger (2000) character and temperament scales require refined clinical skills, as does the assessment of all DSM IV axis 1 diagnoses. For this reason, and to limit measurement error, each prisoner was interviewed by a pair of interviewers (one junior and one senior; see Section 1.3). Before the consensus session, each clinician had to give his (her) own list of diagnoses. From these two series, it is possible to get an idea of the measurement error arising from the interviewer's subjectivity. For instance, concerning the difficult diagnosis of schizophrenia, Cohen's kappa is obtained from the function ckappa() in the library "psy":

❶   ❷
```
> ckappa(mhp.ira[, c("scz.jun", "scz.sen")])
$table❸
     0    1
0  715   11
1   30   43

$kappa
[1] 0.6499695❹
```

The two variables corresponding to the diagnoses given by the two interviewers are scz.jun❶ (diagnosis of schizophrenia (1-yes/0-no) proposed by the junior clinician) and scz.sen❷ (idem for the senior clinician). A two-by-two table❸ gives the distribution of concordance and discordance between them. For 715 prisoners, they both agree that the diagnosis is absent; for 43 prisoners, they both agree that the diagnosis is present; and for 41 prisoners, they disagree. As mentioned previously, it is important to verify that the two frequencies of diagnosis are not very distant from one another. This is the case here: (43 + 11)/799 = 6.8% for the junior clinician and (43 + 30)/799 = 9.1% for the senior clinician. The kappa is 0.65❹. This means that inter-rater agreement is twice as close to perfect agreement as it is to agreement occurring by chance. Some references propose thresholds that could correspond to a high, moderate, or low kappas (Landis and Koch 1977). It is doubtful that they have any real value, as a "good" inter-rater agreement must take into consideration the objective of the measurement process. Some authors have used simulations to show the impact of a moderate inter-rater agreement on the statistical power of a study. Under specific hypotheses, Seigel, Podgor, and Remaley (1992) show that a sample size of 1,000 with a perfectly reliable measurement is equivalent, in terms of statistical power, to a sample size of 2,500 with a measurement obtained with interviewers registering a kappa of 0.6.

Now, if we consider the diagnosis of "post-traumatic stress disorder," it is possible to look for inter-rater agreement with the diagnoses from the junior practitioner, the senior practitioner, and the M.I.N.I. (the structured interview; see Section 1.3). The function lkappa() in the package "psy" is now used:

```
> lkappa(mhp.ira[, c("ptsd.jun", "ptsd.sen", "ptsd.mini")])
[1] 0.4493614
```

The result is not as good. We will see later how to go further in the interpretation of this somewhat poor inter-rater agreement.

The variable "gravity" (rated from 1 to 7) is an ordered categorical variable. If Cohen's kappa (1960) is used to assess inter-rater agreement in this case, the result is

```
> ckappa(mhp.ira[, c("grav.jun", "grav.sen")])
$table
        1     2     3     4     5     6   7
1     101   11❶    5     1    1❷    0   0
2       5   105    16     4     1     1   0
3       1    13    68    29     3     0   0
4       2     1    13   118    33     1   0
5       1     1     3    39   111    14   0
6       0     0     0     2    22    44   6
7       0     0     0     1     5     6   8

$kappa
[1] 0.6337358❸
```

The kappa is 0.63❸. However, this value is difficult to interpret because Cohen's kappa considers all forms of disagreement equally: When the junior physician rates "1" and the senior physician "2"❶, the level of disagreement is equivalent to the situation where the junior physician rates "1" and the senior physician "5"❷. Obviously, this does not make sense. A "weighted" kappa is more adapted to the situation of ordered categorical variables. Here, the level of disagreement will be lower for situation ❶ than for situation ❷. The function wkappa() in the package "psy" can estimate a weighted kappa:

```
> wkappa(mhp.ira[, c("grav.jun", "grav.sen")])
$table
      1     2    3    4    5    6   7
1   101    11    5    1    1    0   0
2     5   105   16    4    1    1   0
3     1    13   68   29    3    0   0
4     2     1   13  118   33    1   0
5     1     1    3   39  111   14   0
6     0     0    0    2   22   44   6
7     0     0    0    1    5    6   8

$weights
[1] "squared"❹

$kappa
[1] 0.9033297❺
```

The result is much better in ❺. The option "squared"❹ is used by default; it formalizes numerically the fact that situation ❶ corresponds to a level of disagreement that is lower than in situation ❷.

If the intra-class correlation coefficient is specifically adapted to the case where the variables are continuous (Conger 1980), it can also be used when an ordered categorical variable has a large number of levels. If we consider "grav.sen" and "grav.jun," then the intra-class correlation coefficient can be obtained from the function icc() in the package "psy":

```
> icc(mhp.ira[, c("grav.jun", "grav.sen")])
$nb.subjects
[1] 796

$nb.raters
[1] 2

$subject.variance
[1] 2.403695❻

$rater.variance
[1] -0.0002275529❼
```

```
$residual
[1] 0.2571371❽

$icc.consistency
[1] 0.9033621

$icc.agreement
[1] 0.9034394❾
```

The total variance of the "gravity" score is broken down into a "subject" variance❻ (difference in gravity from one prisoner to another), a "rater" variance❼ (the systematic discrepancy from one interviewer to another, very small here), and the residual variance❽. The intra-class correlation coefficient❾ of 0.9034 is very close indeed to the weighted correlation coefficient 0.9033 ❺ (in fact, they are asymptotically equal).

Is it now possible to obtain a global picture of inter-interviewer agreement on a qualitative rather than a quantitative basis? It is, for example, using multidimensional exploratory methods as presented in Section 3.6, 3.7, or 3.8. If we consider the diagnoses of schizophrenia ("scz"), post-traumatic stress disorder ("ptsd"), depression ("dep"), generalized anxiety disorder ("gad"), substance abuse or dependence ("subst"), and alcohol abuse or dependence ("alc"), then the function sphpca() in the package "psy" gives

```
> ira <- c("scz.jun", "scz.sen", "psychosis.mini", "ptsd.jun",
    "ptsd.sen", "ptsd.mini", "dep.jun", "dep.sen", "dep.mini",
    "subst.jun", "subst.sen", "subst.mini", "gad.jun",
    "gad.sen", "gad.mini", "alc.jun", "alc.sen", "alc.mini")
> sphpca(mhp.ira[, ira])
```

Clearly, concerning "generalized anxiety disorder," "substance abuse or dependence disorder," "alcohol abuse or dependence," and "depression," there is good inter-rater agreement because the corresponding points are close to one another (Figure 7.13). For "post-traumatic stress disorder," note that the diagnosis proposed by the M.I.N.I.❶ appears closer to the diagnosis of depression proposed by the junior and senior practitioners. Conversely, the diagnosis of PTSD proposed by the junior and senior practitioners❷ seems close to the diagnosis of schizophrenia or psychosis. Perhaps this is an indication that the underlying definition of PTSD for a French clinician differs from the DSM IV definition as operationalized in the M.I.N.I.

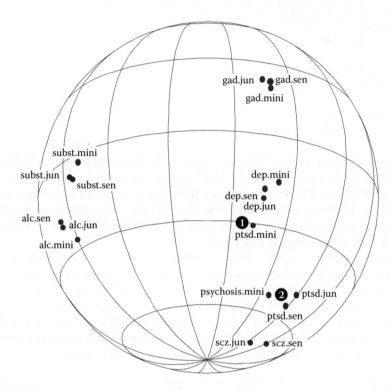

FIGURE 7.13
Spherical representation of the correlation matrix of six diagnoses obtained from three sources (junior clinician, senior clinician, M.I.N.I.).

8

Introduction to Structural Equation Modelling

As seen in Chapters 5 and 6, statistical modelling consists in general of relating an outcome (a dependent variable) to a series of predictors (sometimes also called explanatory variables). This dichotomy is so classic that it appears natural and we sometimes forget that in practice there can be two or three relevant outcomes that could be advantageously studied all together, or that a given predictor can be considered the outcome of another predictor. For instance, the "income" of a young adult can be explained in a linear regression model by the two explanatory variables "educational level" and "parental income." But, at the same time, "educational level" can be explained by "parental income" (and not the reverse, due to a question of temporality). Unfortunately, it is not possible at the moment to implement? in a linear regression model an asymmetrical relationship of this sort between two predictors.

Structural equation modelling deals with this issue and with some others. It can be considered a generalization of linear regression:

1. Multiple outcomes are possible.
2. In a given regression model, a predictor can be explained by another one.
3. Latent, unobserved variables can be introduced just as in factor analysis (see Section 7.4).

Concerning the conditions of validity, they consist principally of the normality of the outcome variable(s).

8.1 Linear Regression as a Particular Instance of Structural Equation Modelling

In a few words: During brainstorming sessions intended to design a statistical analysis plan, it is often useful, on a sheet of paper, to graphically represent the relationships between the variables of interest using boxes and arrows.

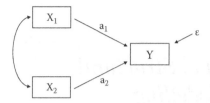

FIGURE 8.1
Graphical representation of the equation $Y = a_0 + a_1X_1 + a_2X_2 + \varepsilon$ following the structural equation modelling paradigm. Y is explained by X_1 and X_2. The double arrow between X_1 and X_2 represents the fact that predictors are freely correlated in a regression model.

Structural equation modelling is based on this perspective (Wright 1921). Following this approach, the linear regression model

$$Y = a_0 + a_1X_1 + a_2X_2 + \varepsilon$$

is represented in the diagram in Figure 8.1. In this diagram, the one-ended arrows that associate X_1 to Y and X_2 to Y symbolize the fact that Y is "explained" or "predicted" by X_1 and X_2 (*). The double-ended arrow that associates X_1 and X_2 reflects the fact that predictors are freely correlated in a regression model.

In Practice: Let us imagine that, in the MHP study, the aim is to model the prisoners' risk of suicidality from (1) the character score in the Cloninger (2000) instrument (this score reflects the level of an underlying personality disorder) and (2) the level of "adversity" during childhood (e.g., being abused, separated from parents, etc.). The idea is that these two explanatory variables are likely associated, and it would be interesting to determine whether both of them are specifically associated with suicidality. The outcome ("suicide.s") will be obtained here from the summation of the questions related to suicide in the structured psychiatric interview M.I.N.I.; its normality should be assessed before any computation, and this part of the analysis will be dealt with in Section 8.3. The Cloninger character score (variable "char.s") will be obtained, as in the previous chapter, from the summation of the 20 character items. The variable "adversity during childhood" is unfortunately not directly available. There are, however, three binary variables related to it: placement during childhood ("outplaced"), ill-treatment during childhood ("abuse"), and separation from parents for longer than six months ("separation"). A summation of these three variables (coded in (yes 1/no 0)) will thus be considered a surrogate for the level of "childhood adversity" for each prisoner.

---

* These arrows are sometimes interpreted as a causal relationship between Y and X_1 or X_2. This is actually going too far, as causality, in observational studies (and most questionnaire surveys are observational studies), can be stated in only very special circumstances (Hill, A. B. 1965). The environment and disease: association or causation?, Proceedings of the Royal Society of Medicine, 58: 295–300.)

The function lm() can then be used to estimate the linear regression model suicide.s = a_0 + a_1 char.s + a_2 child.adv + ε:

❶
```
> mhp.mod$child.adv <- mhp.mod$out.placed + mhp.mod$separation
  + mhp.mod$abuse
                              ❷
> mhp.suirl <- data.frame(scale(mhp.mod[, c("child.adv",
  "char.s", "suicide.s")]))
> mod <- lm(suicide.s ~ char.s + child.adv, data = mhp.suirl)
> summary(mod)

Call:
lm(formula = suicide.s ~ char.s + child.adv, data = mhp.suirl)

Residuals:
  Min       1Q     Median      3Q       Max
-1.6763   -0.5685  -0.2506   0.1187   3.5271

Coefficients:
              Estimate    Std. Error   t value   Pr(>|t|)
(Intercept)   0.008056    0.037365      0.216    0.8294
char.s        0.380401❸   0.038042     10.000    <2e-16    ***❹
child.adv     0.074962❺   0.037917      1.977    0.0485    *  ❻
---
Signif. codes: 0 '***' 0.001 '**' 0.01 '*' 0.05 '.' 0.1 ' ' 1

Residual standard error: 0.9317 on 620 degrees of freedom
  (176 observations deleted due to missingness)
Multiple R-squared: 0.1616, Adjusted R-squared: 0.1589
F-statistic: 59.74 on 2 and 620 DF, p-value: < 2.2e-16
```

The variable "child.adv" is estimated first ❶. Even if they are regularly criticized (see Section 6.5 and Loehlin (1987)), standardized regression coefficients are often used with structural equation models. For this reason, the three variables are standardized (converted so that their mean is equal to 0 and their variance to 1) ❷. In ❸ and ❺ we note the coefficients a_1 and a_2. They are significant at the 5% level in ❹ and ❻.

Let us now see the estimation of a structural equation modelling corresponding to Figure 8.2. The function sem() in the package "sem" can be used:

```
> library(sem)
> N <- dim(na.omit(mhp.mod[, c("child.adv", "suicide.s",
  "char.s")]))[1]
> mhp.cov <- cov(mhp.mod[, c("child.adv", "suicide.s",
  "char.s")], use = "complete.obs")
```

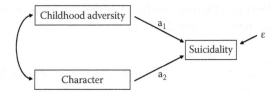

FIGURE 8.2
Structural equation modelling pathway corresponding to the linear regression model where
"suicidality" is the outcome, and "childhood adversity" and "character" are the predictors.

```
> mhp.sem.model <- specify.model()
                    ❶                    ❷     ❸
      child.adv    ->    suicide.s,     a1,    NA
      char.s       ->    suicide.s,     a2,    NA
                    ❹
   suicide.s     <->    suicide.s,     eps,   NA
❺
> mod.sem <- sem(mhp.sem.model, mhp.cov, N,
         ❻
   fixed.x = c("child.adv", "char.s"))
> summary(mod.sem)

   Model Chisquare = -2.7622e-13 Df = 0 Pr(>Chisq) = NA
   Chisquare (null model) = 143.94 Df = 3
   Goodness-of-fit index = 1
   BIC = -2.7622e-13


Normalized Residuals
    Min.     1st Qu.    Median     Mean     3rd Qu.     Max.
-9.23e-16   0.00e+00  0.00e+00  -2.05e-16  0.00e+00  0.00e+00


Parameter Estimates
    Estimate Std Error z value  Pr(>|z|)
a1   0.10732  0.054199  1.9802  0.047682❼ suicide.s <--- child.adv
a2   0.14092  0.014070 10.0157  0.000000  suicide.s <--- char.s
eps  1.78320  0.101136 17.6317  0.000000  suicide.s <--> suicide.s

Iterations = 0
> std.coef(mod.sem)
        Std. Estimate
1 a1    0.074735738❽ suicide.s <--- child.adv
2 a2    0.378011179❾ suicide.s <--- char.s
3 eps   0.838430110❿ suicide.s <--> suicide.s
4       1.00000000  child.adv <--> child.adv
5       0.23170943  char.s    <--> child.adv
6       1.00000000  char.s    <--> char.s
```

After an estimation of the sample size "N" and the correlation matrix "mhp. cov" for the three variables "child.adv," "char.s," and "suicide.s," the model is made explicit using the function specify.model(). This step is not absolutely straightforward. In ❶, it specifies that there is an arrow from "child.adv" to "suicide.s"; in ❷, it specifies that the coefficient associated with this arrow will be called a1; in ❸, the "NA" is used to state that a1 is to be estimated. The residual "ε" associated with "suicide.s" is indicated using a convention: a two-ended arrow between "suicide.s" and "suicide.s" ❹. Notice the blank line❺ that is required.

The function sem() has four arguments: the model obtained from specify.model(), the covariance matrix, the sample size, and the names of the variables that have no residuals❻ (i.e., the variables that have no predictors). This last argument states that the variances and covariance of "child adv" and "char.s" are estimated from the dataset. Note that the covariance corresponds to the two-ended arrow in Figure 8.1.

The results are obtained from the function summary() and the standardized coefficients from std.coef(). The estimations of a_1 and a_2 are, respectively, 0.075 and 0.38; these results are identical to those obtained with the linear regression model (❸ and ❺ in the previous output). The p-value of the test $a_1 = 0$ is obtained in ❼. The variance of the residual "ε" is in ❿.

8.2 Factor Analysis as a Particular Instance of Structural Equation Modelling

In a few words: Along the lines of the previous section, it is possible to represent a factor analysis model with boxes corresponding to variables and arrows corresponding to relationships between outcome and predictors. Let us consider, for reasons of simplicity, a one-factor analysis model estimated from three observed items X_1, X_2, and X_3 with mean zero:

$$X_1 = a_1F + \varepsilon_1$$

$$X_2 = a_2F + \varepsilon_2$$

$$X_3 = a_3F + \varepsilon_3$$

Three arrows point from F, the latent unobserved variable, to X_1, X_2, and X_3 so that Figure 8.3 is obtained.

In Practice: In the previous section, we derived the variable "childhood adversity" from the summation of three apparently related binary variables. Without any supportive psychometric analysis, a summation of this type is open to

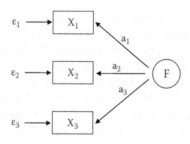

FIGURE 8.3
Graphical representation of the one-factor analysis model $X_1 = a_1F + \varepsilon_1$, $X_2 = a_2F + \varepsilon_2$, $X_3 = a_3F + \varepsilon_3$. One-ended arrows relate the predictor F to the outcomes X_1, X_2, X_3. To differentiate F (a latent variable) from the other observed variables, it is represented in a circle instead of a square.

discussion. Indeed, as seen in Chapter 7, a composite score should be computed only if the items form a unidimensional set, and adequate unidimensionality is observed if the items verify a one-factor factor analysis model with substantial loadings. Using the function `factanal()` we obtain the following:

```
> factanal(na.omit(mhp.mod[, c("out.placed", "separation",
  "abuse")]), factors = 1❶)

Call:
factanal(x = na.omit(mhp.mod[, c("out.placed", "separation",
  "abuse")]), factors = 1)

Uniquenesses:
out.placed  separation  abuse
   0.648       0.665     0.745

Loadings:
             Factor1
out.placed    0.594❷
separation    0.579❸
abuse         0.505❹

               Factor1
SS loadings     0.942
Proportion Var  0.314
```

The degrees of freedom for the model is 0❺ and the fit was 0

A one-factor model is requested in ❶. The three loadings ❷, ❸, and ❹ are above 0.5; this supports the computation of the composite score obtained from the summation of the three corresponding items. The meaning of this score can be obtained from "what the three items have in common"; here, the label "adverse childhood" has been retained. Curiously, the degree of

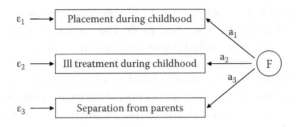

FIGURE 8.4
Structural equation modelling pathway corresponding to a one-factor factor analysis model. The closer that a_1, a_2, and a_3 are to 1 (or –1), the more robust and well-identified the latent variable F.

freedom of the model is 0 ❺. This is because the number of coefficients to estimate (here three) is just equal to the number of correlation coefficients available to estimate the model (here the correlations between "out placed," "separation," and "abused"). The model fit is thus (artificially) perfect. For four or more variables, this will no longer be the case.

Let us see now what is provided by the sem() function applied to the diagram presented in Figure 8.4:

```
> N <- dim(na.omit(mhp.mod[, c("out.placed", "separation",
  "abuse")]))[1]
> mhp.cov <- cov(mhp.mod[, c("out.placed", "separation",
  "abuse")], use = "complete.obs")
> mhp.sem.model <- specify.model()
         ❶
    child.adv  ❷-> out.placed, a1,   NA
    child.adv  ❷-> abuse,      a2,   NA
    child.adv  ❷-> separation, a3,   NA
    child.adv  <-> child.adv,  NA,   1❸
    out.placed <-> out.placed, eps1, NA
    abuse      <-> abuse,      eps2, NA
    separation <-> separation, eps3, NA

Read 7 records
         ❹
> mod.sem <- sem(mhp.sem.model, mhp.cov, N)
> std.coef(mod.sem)
         Std. Estimate
a1       0.5935872❺     out.placed <--- child.adv
a2       0.5045525❻         abuse <--- child.adv
a3       0.5791693❼    separation <--- child.adv
         1.0000000       child.adv <--> child.adv
eps1     0.6476542     out.placed <--> out.placed
eps2     0.7454268          abuse <--> abuse
eps3     0.6645629     separation <--> separation
```

The latent variable F is labeled "child.adv"❶. The arrows from "F" to the three observed items are specified in ❷. The variance of the latent variable is fixed to a prespecified value because it is neither observed nor explained by any variable. It is fixed at 1 here❸. Because all observed variables are outcomes, there is no argument fixed.x = in the call on the sem function sem(). The three standardized coefficients are in ❺, ❻, and ❼. They are the same as those found in the output of the one-factor factor analysis model (❷, ❸, ❹).

There is, however, a potential problem with the model developed in the present section. The three variables "placement during childhood," "ill treatment during childhood," and "separation from parents" are all dependent variables and should therefore have a normal distribution. This is definitely not the case here, as these three variables are all binary. It can be noted that we had the same problem with factor analysis in the previous section. We argued there that the technique is basically exploratory and thus less sensitive to conditions of validity. This argument is less relevant here because structural equation modelling obviously has inferential ambitions (the test that a path coefficient is different from 0 can be of real value). In all cases, results should be considered very cautiously. At the end of the next section we will see that certain techniques have been proposed to deal with this issue.

8.3 Structural Equation Modelling in Practice

In a few words: Three important steps can be defined concerning the design of a structural equation model:

1. The choice of the variables of interest (observed and latent);
2. The determination of the pattern of relationships between these variables; and
3. The assessment of model fit and parsimony.

Concerning the choice of the variables of interest, there are two central considerations:

1. Outcome variables are assumed to follow a normal distribution, and measurement error should be as small as possible. We will see below that the condition of normality for the outcome variables can theoretically be relaxed; it is however by far preferable for the condition to be met, in order to ensure that the convergence process leads to reliable estimated parameters. In practice, normality can be verified

graphically as has been explained previously (see, for example, Section 4.7 and Section 5.1).

2. Variables that are likely to measure the same construct (e.g., "father's income," "mother's income," "number of rooms in the house," etc.) should be put together in a latent variable (here, socio-economic status). The latent variable will be reliable if the underlying observed variables form a unidimensional set (this can be verified with the scree plot) and have good internal consistency (which can be estimated with the Cronbach alpha). When only two, three, or four items dealing with the same notion are grouped, these psychometric properties can be apprehended in a simpler manner using inter-item correlations, which should be fairly homogeneous and sufficiently high (e.g., above 0.3).

Concerning the pattern of relationships between the variables of interest, it should be carefully considered *a priori*, on the basis of data in the literature or according to a theoretical rationale. There is often a temptation to draw a huge number of arrows representing all the hypotheses that arise at one moment or another in the researcher's mind; this is in general a bad idea. Experience shows that a simple problem represented by a small number of variables and arrows provides stable results that are usually easier to interpret.

Finally, it is also recommended to use a pattern of arrows with no loops (Cox and Wermuth 1996) (Figure 8.5).

When the model has been designed, the coefficients associated with each arrow can be numerically determined. In R this will be classically obtained

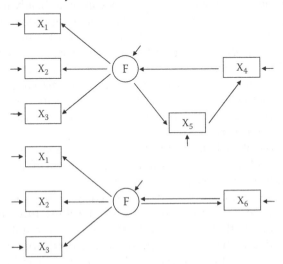

FIGURE 8.5
In a structural equation model, "loops" should be avoided (as between F, X_5, and X_4, or between F and X_6).

from the function sem() in the package "sem" (the more recent package "lavaan" could become a preferable alternative in the near future). Model goodness-of-fit is traditionally assessed from a series of indices and statistical tests of hypothesis.

We have already discussed the difficulty in interpreting results from goodness-of-fit tests (Section 5.2): If the number of patients is extremely large, the null hypothesis that the model is acceptable will always be rejected. In practice, these tests can be useful in at least two situations: (1) if the sample size is large (e.g., thousands or more) and the p-value is above 0.05 (in which case the model fit is obviously good), and (2) if the sample size is small (e.g., 100 or 200 or less) and the p-value is lower than 0.05 (in which case the model fit is likely to be bad).

Among the long list of fit indices available, two are widely used: the RMSEA (root mean square error of approximation) and the NFI (normed fit index). A perfect fit corresponds to an RMSEA of 0 and an NFI of 1. An RMSEA smaller than 0.05, or even 0.08, is generally considered a good fit, while an RMSEA higher than 0.1 corresponds to a poor fit (MacCallum, Browne, and Sugawara 1996; Hu and Bentler 1999). As far as the NFI is concerned, a value higher than 0.95 is generally considered a good fit, while a value lower than 0.9 corresponds to a poor fit.

It happens frequently that several models are eligible to represent the pattern of correlations between the variables with an acceptable fit. These models differ by the fact that some have a small number of additional arrows. These supplementary arrows are potentially useful because model fit is necessarily better and because a larger number of pathways are numerically accessible. On the other hand, more arrows can be a disadvantage because the models become more complex, and less easy to apprehend, and thus in the end they prove less informative. To solve this dilemma of a necessary balance between goodness-of-fit and simplicity, certain parsimony indices can be used. These indices have no intrinsic meaning; they are useful only for comparing several models: the one with the lowest value has the most favorable ratio between goodness-of-fit and simplicity. Two well-known parsimony indices are the AIC (Akaike information criterion) and the BIC (Bayesian information criterion).

While the choice of variables of interest, the determination of the pattern of arrows that connect these variables, and the assessment of model fit and parsimony are the three main steps in structural equation modelling, there is a fourth point that should be considered here: The possible use of non-normal variables and more generally the question of robust inferences. Indeed, it frequently happens that binary questions could be profitably included in a structural model. If these binary questions are predictors (i.e., if they are not on the receiving end of any arrow), then the model can be estimated as usual. If they are outcomes (or both outcomes and predictors), the conditions of validity are not fulfilled and the results are unreliable: p-values and confidence intervals should be considered as mere indications. To deal with

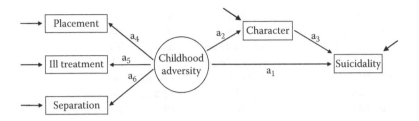

FIGURE 8.6
Structural equation modelling relating childhood adversity and character to suicidality. As opposed to Figure 8.2, "character" and "childhood adversity" do not have a symmetrical role; there can be a direct relationship between childhood adversity and suicidality and an indirect relationship mediated via "character." In addition, "childhood adversity" is no longer a composite score; it is now a latent variable obtained from the three items that make up the composite score.

this problem, the use of a modified version of the correlation matrix has been suggested, using, instead of the classic Pearson correlations, the so-called tetrachoric or bi-serial correlations. The function hetcor() in the package "polycor" can be used for this. More generally, if there is doubt concerning the truth of the normality assumption, it is possible to use bootstrap estimates of p-values and confidence intervals (see Section 6.7)

In Practice: Section 8.1 focused on a structural equation model that was equivalent to a linear regression model where the prisoner's level of suicidality was the outcome, and the two variables Cloninger's (2000) "character" score and "level of childhood adversity" were the predictors. "Level of childhood adversity" was a composite score obtained from the summation of three binary variables: placement during childhood, ill-treatment during childhood, and separation from parents longer than six months.

It is possible to use the flexibility of structural modelling to improve this model. First, the two predictors are not symmetrical: Childhood adversity could "cause" a low score for character, which, in turn, could be related to a high level of suicidality. Second, it would be better to consider "childhood adversity" as a latent variable obtained from the three binary indicators "out-placed," "abuse," and "separation" rather than a composite score. The model integrating these two observations is presented in Figure 8.6

It is possible to estimate the coefficients of the model corresponding to Figure 8.6 using the sem() procedure:

```
> mhp.int <- na.omit(mhp.mod[, c("out.placed", "separation",
  "abuse", "suicide.s", "char.s")])
> N <- dim(mhp.int)[1]
> mhp.cov <- cov(mhp.int)
> mhp.sem.model <- specify.model()
      child.adv      ->   out.placed,   a4,    NA
      child.adv      ->   separation,   a6,    NA
```

```
        child.adv      ->   abuse,        a5,     NA
        child.adv      ->   char.s,       a2,     NA
        child.adv      ->   suicide.s,    a1,     NA
        char.s         ->   suicide.s,    a3,     NA
        char.s         <->  char.s,       eps1,   NA
        suicide.s      <->  suicide.s,    eps2,   NA
        child.adv      <->  child.adv,    NA,     1
        out.placed     <->  out.placed,   eps3,   NA
        separation     <->  separation,   eps4,   NA
        abuse          <->  abuse,        eps5,   NA
>  mod.sem <- sem(mhp.sem.model, mhp.cov, N)
>  summary(mod.sem)
                                                          ❶
   Model Chisquare = 3.4028  Df = 4 Pr(>Chisq) = 0.49281
   Chisquare (null model) = 317.36 Df = 10
   Goodness-of-fit index = 0.99783
   Adjusted goodness-of-fit index = 0.99185
   RMSEA index = 0❷ 90% CI: (NA, 0.056364)
   Bentler-Bonnett NFI = 0.98928❸
   Tucker-Lewis NNFI = 1.0049
   Bentler CFI = 1
   SRMR = 0.015486
   BIC = -22.335❹

Normalized Residuals
   Min.      1st Qu.    Median     Mean      3rd Qu.     Max.
 -7.73e-01  -6.51e-02  8.59e-06  2.05e-02  1.05e-01  1.14e+00

Parameter Estimates
        Estimate Std Error  z value   Pr(>|z|)
a4      0.26088   0.023244  11.2234 0.0000e+00  out.placed <--- child.adv
a5      0.28802   0.026564  10.8425 0.0000e+00  separation <--- child.adv
a6      0.22429   0.023602   9.5027 0.0000e+00  abuse      <--- child.adv
a2      1.19763   0.199858   5.9924 2.0673e-09  char.s     <--- child.adv
a1      0.13936   0.075291   1.8510 6.4171e-02❺ suicide.s  <--- child.adv
a3      0.13647   0.014939   9.1350 0.0000e+00  suicide.s  <--- char.s
eps1   13.87005   0.844666  16.4207 0.0000e+00  char.s     <--> char.s
eps2    1.77684   0.101563  17.4949 0.0000e+00  suicide.s  <--> suicide.s
eps3    0.11166   0.011123  10.0391 0.0000e+00  out.placed <--> out.placed
eps4    0.16419   0.014543  11.2904 0.0000e+00  separation <--> separation
eps5    0.15749   0.011534  13.6539 0.0000e+00  abuse      <--> abuse

Iterations = 54
```

The first part of the listing is similar to what was discussed previously. The p-value of the goodness-of-fit test is in ❶. It is above 0.05, which is in favor of an acceptable fit. The RMSEA is equal to 0❷; this is also in favor of a good fit

(<0.05); and the same is true for the NFI at 0.99❸ and thus close to 1, and in particular above 0.95 (which is a classic threshold for a good fit).

The Bayesian Information Criterion (BIC) is −22.3❹. This index can be used to compare the balance between fit and simplicity in a series of models. For example, if some arrows were added to or removed from Figure 8.6, the new model and the old one could be compared on the basis of their BIC. If the new model has a BIC lower than −22.3, the new model is likely better. Of course, if the new BIC is just slightly below −22.3 (e.g., −22.5), the corresponding model will be only marginally better. It has been suggested that a decrease of 5 points in BIC must be observed to claim "strong evidence" that the new model is indeed better (this would correspond here to a new BIC below −27.3); for a decrease as large as 10 points, this would correspond to "conclusive evidence" of a better model (Raftery 1993).

All the arrows have coefficients that are statistically significant at the 5% level, all except the arrow between childhood adversity and suicidality❺.

The standardized coefficients can be obtained using the function std.coef():

```
> std.coef(mod.sem)
         Std. Estiamte
a4       0.6153706      out.placed    <---    child.adv
a5       0.5793544      separation    <---    child.adv
a6       0.4920293      abuse         <---    child.adv
a2       0.3061371❻     char.s        <---    child.adv
a1       0.0955606❼     suicide.s     <---    child.adv
a3       0.3660729❽     suicide.s     <---    char.s
eps1     0.9062801      char.s        <-->    char.s
eps2     0.8354401      suicide.s     <-->    suicide.s
         1.0000000      child.adv     <-->    child.adv
eps3     0.6213190      out.placed    <-->    out.placed
eps4     0.6643485      separation    <-->    separation
eps5     0.7579072      abuse         <-->    abuse
```

The direct relationship between childhood adversity and suicidality yields a standardized coefficient of 0.096❼. The indirect relationship between the two variables mediated by the character score is composed of a first path with a standardized coefficient of 0.306❻ and a second path with a standardized coefficient of 0.366❽. The "strength" of this indirect relationship can be obtained from the product of the two coefficients (that is, 0.306 × 0.366 = 0.112). This result is comparable even if it is a little higher than the direct relationship.

Once a first model has been estimated, it may need to be modified, either because some pathways have coefficients close to zero and the corresponding arrows could therefore be removed, or because the fit is poor and in this case some arrows should be added. Such post hoc modifications of a structural equation model, if they are widely implemented, raise problems that have been already encountered in stepwise regression modelling (see Section 6.3); the approach is based more on an exploratory perspective rather than a purely confirmatory, hypothetico-deductive perspective. In practice, the function mod.indices() can be used to determine which arrows could be added:

```
> mod.indices(mod.sem)

5 largest modification indices, A matrix:
    abuse:suicide.s        suicide.s:abuse      separation:suicide.s
       1.9837015❾            1.5256669                1.1401154
suicide.s:separation     separation:char.s
       0.5040286             0.2444045


5 largest modification indices, P matrix:
    suicide.s:abuse       suicide.s:separation    char.s:separation
       2.2978599❿            0.9441416               0.2781088
    abuse:out.placed      char.s:out.placed
       0.2158709             0.1893293
```

The indices follow a one-degree of freedom, chi-square statistic so that values greater than 4 are statistically significant at the 5% level, and higher values correspond to stronger relationships. New arrows should not be added on the sole basis of statistical considerations but also—and perhaps above all—on scientific grounds. Here, no extra pathway seems of interest. The one-headed arrows are presented first (A matrix). The pathway "abuse -> suicide.s"❾ appears to be the most meaningful, but the chi-square value of 1.98 is too small for consideration. The double-headed arrows are presented after (P matrix); here again, the pathway "abuse <-> suicide.s"❿ appears to be the most meaningful, but the chi-square is too small for consideration.

Now, one problem in the model presented in Figure 8.6 is that the three variables "placement during childhood," "ill-treatment during childhood," and "separation from parents longer than six months" are binary variables and thus do not follow the normal distribution that is required for all outcomes present in a structural equation model. One way to deal with this issue is to consider that these variables are, in fact, continuous and that they have been artificially dichotomized due to the choice of a yes/no response pattern. This is typically the case for "ill-treatment during childhood" (there may be different levels of ill-treatment); it is less clear for the other two variables. The function hetcor() in the package "polycor" can estimate the Pearson correlations that would have been observed if the underlying continuous variables had been collected instead of the yes/no answers. This correlation matrix can then be obtained using the sem() function to obtain more satisfactory coefficients and p-value estimates:

```
> mhp.polycor <- mhp.int
                              ❶
> mhp.polycor$out.placed <- as.ordered(mhp.polycor$out.placed)
> mhp.polycor$separation <- as.ordered(mhp.polycor$separation)
> mhp.polycor$abuse <- as.ordered(mhp.polycor$abuse)
> mhp.cov <- hetcor(mhp.polycor)
> mhp.sem.model <- specify.model()
        child.adv    ->    out.placed,   a4,    NA
        child.adv    ->    separation,   a6,    NA
        child.adv    ->    abuse,        a5,    NA
        child.adv    ->    char.s,       a2,    NA
        child.adv    ->    suicide.s,    a1,    NA
        char.s       ->    suicide.s,    a3,    NA
        char.s       <->   char.s,       eps1,  NA
        suicide.s    <->   suicide.s,    eps2,  NA
        child.adv    <->   child.adv,    NA,    1
        out.placed   <->   out.placed,   eps3,  NA
        separation   <->   separation,   eps4,  NA
        abuse        <->   abuse,        eps5,  NA
```

```
> mod.sem <- sem(mhp.sem.model, mhp.cov$correlations, N)
> summary(mod.sem)

  Model Chisquare = 7.4186 Df = 4 Pr(>Chisq) = 0.11535❷
  Chisquare (null model) = 630.91 Df = 10
  Goodness-of-fit index = 0.9953
  Adjusted goodness-of-fit index = 0.9824
  RMSEA index = 0.037068❸ 90% CI: (NA, 0.078244)
  Bentler-Bonnett NFI = 0.98824❹
  Tucker-Lewis NNFI = 0.98624
  Bentler CFI = 0.9945
  SRMR = 0.020075
  BIC = -18.320
```

```
Normalized Residuals
   Min.      1st Qu.    Median     Mean     3rd Qu.     Max.
-8.97e-01   -1.30e-05  4.95e-06  3.99e-02  8.72e-02  1.47e+00
```

```
Parameter Estimates
       Estimate  Std Error  z value  Pr(>|z|)
a4     0.791914❺  0.041461  19.1003  0.0000e+00 out.placed <--- child.adv
a5     0.740471❻  0.041256  17.9483  0.0000e+00 separation <--- child.adv
a6     0.612979❼  0.041558  14.7499  0.0000e+00 abuse      <--- child.adv
a2     0.310110❽  0.044276   7.0040  2.4878e-12 char.s     <--- child.adv
a1     0.084514❾  0.044450   1.9013  5.7261e-02 suicide.s  <--- child.adv
a3     0.369119❿  0.039238   9.4072  0.0000e+00 suicide.s  <--- char.s
```

```
eps1  0.903833    0.052851  17.1014  0.0000e+00 char.s       <--> char.s
eps2  0.837260    0.047606  17.5873  0.0000e+00 suicide.s    <--> suicide.s
eps3  0.372872    0.044624   8.3559  0.0000e+00 out.placed   <--> out.placed
eps4  0.451703    0.042779  10.5591  0.0000e+00 separation   <--> separation
eps5  0.624257    0.043440  14.3706  0.0000e+00 abuse        <--> abuse

Iterations = 13
```

The variables "abuse," "out.placed," and "separation" must first be defined as ordered variables❶ so that hetcor() can compute the appropriate Pearson, tetrachoric, or biserial correlation coefficients. The model specifications are identical. Concerning the results, model fit is still good (❷, ❸, ❹). Because a correlation matrix is analysed, instead of a covariance matrix as previously, standardised coefficients are directly provided in the section "Parameter Estimates." It can be noted that the coefficients associated with the binary variables are now higher (❺, ❻, ❼) while the other coefficients (❽, ❾, ❿) are comparable to those obtained previously (see ❻, ❼, ❽ in the listing above). These modifications are classic when switching from Pearson correlations estimated from binary variables to tetrachoric correlations.

In the model in Figure 8.6, there are three observed binary variables and two observed numerical variables (score for character and score for suicidality). We have dealt appropriately with the binary variables. The normality of the two numerical variables should have been verified at the very beginning of the analysis, during the initial description step. While the histogram of "char.s" reveals a distribution that is approximately normal, this is definitely not the case with the score for suicidality, which presents a marked floor effect (Figure 8.7).

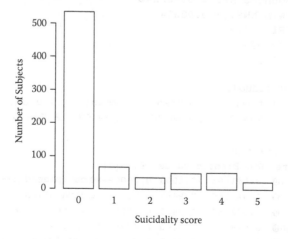

FIGURE 8.7
Histogram of the suicidality score. There is obviously a floor effect giving a non-normal distribution.

Robust estimates of the coefficient confidence intervals would therefore be extremely useful here. They can be obtained using a bootstrap procedure (see Section 6.7). The function boot.sem() in the package "sem" is specifically dedicated to this task:

```
> res.boot <- boot.sem(mhp.polycor, mod.sem, R = 3000,
    cov = hcor)
> summary(res.boot, type = "bca")

Call: boot.sem(data = mhp.polycor, model = mod.sem, R = 3000,
    cov = hcor)

Lower and upper limits are for the 95 percent bca confidence
    interval
```

	Estimate	Bias	Std.Error	Lower	Upper
a4	0.7919141	0.005849382	0.06478325	0.6600242	0.9108422
a5	0.7404706	-0.005536658	0.06034637	0.6342110	0.8740950
a6	0.6129792	-0.001425222	0.06117543	0.4927334	0.7301619
a2	0.3101098	0.009158078	0.05640611	0.1896687	0.4085305
a1	0.0845139	0.013086498	0.05993643	-0.0390772	0.1937796 ❶
a3	0.3691193	-0.005001986	0.04248723	0.2904559	0.4548526
eps1	0.9038327	-0.008945378	0.03611995	0.8330875	0.9640237
eps2	0.8372602	-0.006676272	0.03069648	0.7803455	0.8947387
eps3	0.3728720	-0.013494213	0.10333827	0.1703654	0.5643598
eps4	0.4517030	0.004528595	0.08910252	0.2359085	0.5977704
eps5	0.6242566	-0.001996183	0.07470107	0.4668608	0.7572105

As observed previously, no 95% confidence intervals contain 0, except coefficient $a_1$❶, which corresponds to the direct pathway between childhood adversity and suicidality.

9

Introduction to Data Manipulation Using R

Nowadays, most statistical calculations require less than a few seconds on a CPU (central processor unit) laptop computer. It is thus often considered that people analysing data mainly spend their time in deliberations concerning the methods and routines they should optimally implement. Unfortunately, this is not true. Most of the time is, in fact, simply devoted to data manipulations.

Datasets corresponding to questionnaire studies generally concern hundreds of variables measured on thousands of subjects. These datasets are often obtained from several sources and can be segmented into many parts, each corresponding to a specific section of the questionnaire. Moreover, the labelling of some variables can be incompatible with R (e.g., the variable name "age-of-mother"), while the coding of categorical variables can raise problems (it may be that yes/no questions are coded as 1 for "yes" and 2 for "no," or a "yes"=1 and "no" = 0 coding may be preferred in some circumstances). Many considerations of this sort explain why datasets must be manipulated before or even during the core analysis procedures. Unfortunately, these manipulations are regularly the source of dramatic errors; it is even plausible that critical errors occurring during the analysis of a questionnaire more probably arise from problems in data management rather than from purely statistical errors.

We are now going to focus on the main R functions required for data manipulation. We deal with the question of importing and exporting data, then look at the manipulation of datasets and variables, and conclude with the essential checks for inconsistencies.

9.1 Importing and Exporting Datasets

In a few words: Most often, datasets are provided in a "text" format. Observations are in lines, and variables are in columns. Each value is separated from the next with a comma (","), a semi-colon (";"), or a tabulation. The first line generally gives the names of the variables.

Variable names must comply with certain conditions: no space and no unusual character such as "?", "[", "-", "@", "$", "^", "£", etc. There is an important exception to this second condition: the character "." is authorized. This may be helpful in facilitating an understanding of variable labels such as "father.job" or "history.car.accident," the latter of which could also be labeled "hist.car.acc." A variable name cannot begin with a number (e.g., "3var"), but it can end with a number (e.g., "var1," "var2," …).

Upper- ("A") and lower-case letters ("a") are distinguished, so that the two variables "Age" and "age" are different. This is intended to reduce the probability of mistakes in calling a variable or a function. It is also the reason for numerous, apparently inexplicable bugs. It can thus be suggested that only lower-case letters should be used in variable labelling.

The R functions to import datasets in text format are `read.csv()`, `read.csv2`, `read.delim()`, or `read.delim2()`.

Datasets can also be imported from files obtained from spreadsheet software (such as Excel) or from other statistical software (such as SPSS or Stata with the functions `read.spss()` or `read.dta()` in the package "foreign").

In any case, it is not rare for some errors to occur during the import process. For this reason it is always necessary to have a look at the dataset once it is obtained; the function `str()` can be used for this. For the same reason, it is a good habit to systematically check a series of points that frequently raise problems; for example, variable names should not have been modified during import, numerical variables with decimals should be written with a "." (while decimals can be written "," in certain countries), and variable type (numerical, integer, or factor) should be consistent with the nature of the variable and missing values should be coded with NA.

In Practice: Imagine that the objective is to import the file that contains data collected by the "junior" interviewer in the MHP study. This file is a text file named "junior.csv."

The first step is to determine the name of the R object that will correspond to this dataset (in R terminology, a dataset is called a "data frame"). For simplicity, the name "mhp.j" was chosen here.

The second step is to obtain the pathway that indicates the file location. For Windows users, click on the file icon with the right button of the mouse, and then click "property" and "copy" the pathway.

The third and last step is to write: "mhp.j <- read.csv("" in the script window of R, and then to paste the pathway changing backward slashes ("\") to forward slashes ("/") to obtain, for instance,

```
mhp.j <- read.csv("C:/Users/Bruno/Book AQR/Data MHP/junior.csv")
```

Then click on the icon "run" (❷ in Figure 1.2). The result in the "R control" window should be as follows:

❹
```
> mhp.j <- read.csv("C:/Users/Bruno/Livre AQR/Data SMP/
  junior.csv")
>
```

The instruction read.csv()❹ is designed to import comma (",") separated text files. In some countries, commas replace decimal points and in this case semi-colons (";") are used as separators. For this purpose, the instruction read.csv2() should be used. When variable values are separated by a tab character, the function read.delim() is appropriate, as for read.delim2() when the tab is the separator and decimals are represented by commas.

The instruction above "mhp.j <- read.csv(...)" stores "junior.csv" in the data frame "mhp.j" but does not display the content of this object. This is a good point because a very large dataset would be impossible to put on view on a computer screen. The function str() can be used to take a superficial look of the imported dataset:

```
> str(mhp.j)
                    ❶              ❷
'data.frame': 799 obs. of 122 variables:
       ❸             ❹            ❺
$ centre      : int 1 1 1 1 1 1 1 1 1 1 ...
$ file        : int 121 152 151 122 124 123 125 126 127 128 ...
$ number      : int 1 2 3 4 5 6 7 8 9 10 ...
$ name.junior : Factor w/ 26 levels "Bertrand",..: 4 3 3 4 3
                    4 4 4 4
$ dsm.dep1    : int 1 1 1 1 0 1 1 1 1 0 ...
.... [TRUNCATED]
```

The number of observations ❶ and variables ❷ is provided first, followed by each variable name ❸, with its type ❹ and the first values ❺. The most common variable types are "integer," "numerical" (basically numbers with decimals), and "factors" (categorical variables).

Due to the large audience of the Microsoft spreadsheet program, a dataset may be in Excel format. In this situation, the simplest attitude is probably to convert the file as a "csv" (comma-separated value) file using Excel and then to import the "csv" file into R as described above. This conversion can be done using the "save as" option in Excel in the "File" menu. It is possible, however, to directly import an Excel file into R using the "RODBC" package:

```
> library(RODBC)
          ❷                    ❶
> odbc.junior <- odbcConnectExcel("C:/Users/Bruno/Book AQR/
    Data MHP/juniorXLS972003.xls")
          ❸                          ❹
> sqlTables(odbc.junior)$TABLE_NAME
          ❺
[1] "junior$"
          ❻
> mhp.jxls <- sqlFetch(odbc.junior, sqtable = "junior$")
```

```
> str(mhp.jxls)
'data.frame': 799 obs. of 126 variables:
$ cent : num 1 1 1 1 1 1 1 1 1 1 ...
$ cah  : num 121 152 151 122 124 123 125 126 127 128 ...
$ det  : num 1 2 3 4 5 6 7 8 9 10 ...
$ inv  : Factor w/ 26 levels "
.... [TRUNCATED]
```

The function adbcConnectExcel()❶ is used to determine the connection handle, which is stored in ❷. In an Excel file, there may be several spreadsheets, the object TABLE _ NAME❹ obtained from the function sqlTables()❸ lists the names of all the tables in the Excel file. Here, there is only one spreadsheet named "junior$"❺. Finally, the function sqlFetch() is used to import the Excel file.

It is also possible with R to import a dataset saved using different statistical software. The functions read.spss() and read.dta() from the "foreign" package can be used to read SPSS or Stata files.

It is often necessary to export as well as import datasets. If text format is preferred, the function write.csv() or write.csv2() can be used in the same way as read.csv() or read.csv2(). For instance, in a country where decimal points are written as dots ("."), the export of a "comma-separated values" version of the R data frame "mhp.j" will be obtained via the following instruction:

```
> write.csv(mhp.j, "C:/Users/Bruno/Book AQR/Data MHP/
                 ❶                           ❷
  mhpjunior.csv", row.names = FALSE, quote = FALSE)
```

The file "mhpjunior.csv" is now on the hard disk "C:" in the repertory C:/Users/Bruno/Book AQR/Data MHP/. Two options have been used: ❶ to discard rows numbers that are unnecessary here, and ❷ to eliminate quotes ("") in variable names and factor labels.

Several datasets can be stored together in a single R object using the function save():

```
> save(mhp.ex2, mhp.ex, mhp.mod, file = "C:/Users/Bruno/Book AQR/
  Data MHP/dataAQR")
```

Here, the data frames "mhp," "mhp.ex," and "mhp.mod" are stored in the object "dataAQR". In a new R session, this object can be loaded:

```
> load("C:/Users/Bruno/Book AQR/Data MHP/dataAQR")
```

so that the data frames "mhp.ex2," "mhp.ex," and "mhp.mod" are now all available again.

Finally, all the outputs obtained during an R session can be saved using the command "Save to File..." in the "File" menu. And all the commands that have been used can be saved using the command "Save History..." in the same menu.

9.2 Manipulation of Datasets

In a few words: Some basic commands concerning the manipulation of datasets are essential. How is a subset of subjects selected (e.g., males or people older than 30 years), or a subset of variables? How are two data frames merged (two sets of subjects having the same observations or, conversely, two sets of variables measured on the same subjects)? How is a data frame sorted according to the values of a series of variables?

In Practice: An R data frame is an array with observations in rows and variables in columns. If "mhp" is the name of a data frame, it is possible to extract the value of the third ❷ variable observed for the 243th ❶ subject with the instruction:

```
     ❶   ❷
> mhp[243, 3]
[1] 741
```

More generally, inside the brackets, everything that involves the rows of the data frame is to the left of the comma "," while everything that involves the columns is to the right of the comma. The value of the observation is here "741."

The variable representing the prisoner's identity is named "file." It corresponds to the third column of the data frame "mhp," and it can be extracted using one of the following three instructions:

```
       ❶
> x <- mhp[, 3]
> x <- mhp[, "file"]
> x <- mhp$file
```

There is nothing to the left of the comma "," ❶; this reflects the fact that all observations should be selected.

Along the same lines, if we now want to extract the subset of variables numbers 3, 5, 6, and 10 in the subsamples of patients number 1 to 120 and 125 to 250 and to store this subset in the data frame named mhp.ex3, the instruction is as follows:

```
          ❶              ❷
> mhp.ex3 <- mhp[c(1:120, 125:250), c(3, 5, 6, 10)]
```

The instruction 1:120 stands for the vector of 120 integers "1, 2, ..., 120" while c() combines several observations or variables.

However, it is not recommended to select variables by their rank. The risk of error is too great. Instead, the recommendation is to select variables using their names. For instance, the subset of the "mhp" data frame containing all Cloninger's (2000) temperament variables could be obtained using the following instructions:

```
> temp <- c("ns1", "ns2", "ns3", "ns4", "ns5", "ha1", "ha2",
  "ha3", "ha4", "ha5", "rd1", "rd2", "rd3", "rd4", "rd5")
> mhp.temp <- mhp[, temp]
```

Note that the syntax mhp[, temp] is used here and not mhp[, "temp"] because temp is not a variable name in the dataset mhp but rather a vector that contains a series of variable names in mhp.

The subset of temperament variables measured on prisoners older than 30 years can be obtained by

```
> mhp.old.temp <- mhp[mhp$age > 30, temp]
```

Now the two subsets of temperament variables measured in prisoners presenting (depression and schizophrenia) or (depression or schizophrenia) according to the junior interviewer can be obtained using the instruction:

```
> mhp.depandscz.temp <- mhp[mhp$dep.jun == 1 & mhp$scz.jun == 1,
  temp]
> mhp.deporscz.temp <- mhp[mhp$dep.jun == 1 | mhp$scz.jun == 1,
  temp]
```

It is notable here that a double equal "==" is used. The symbols for "below or equal to" and "not equal" are, respectively, <= and !=.

The function subset() is perhaps the neatest way to extract subsets from a data frame. Thus, temperament variables of prisoners presenting depression and schizophrenia can be obtained using the following instruction:

```
                                          ❶
> mhp.depandscz.temp <- subset(mhp, dep.jun == 1 & scz.jun == 1,
        ❷
  select = temp)
```

The first option ❶ involves the selection of observations, the second ❷ involves the selection of variables. An interesting use of subsets concerns situations where a series of variables are to be discarded. For instance, the subset of the data frame "mhp" discarding the five temperament variables related to novelty-seeking could be obtained from the following instruction:

```
> mhp.depandscz.temp <- subset(mhp, select = c(-ns1, -ns2,
  -ns3, -ns4, -ns5))
```

Now instead of subsetting data frames, how do we go about merging them? Merging is often necessary when various parts of a questionnaire have been captured and stored in different places. The function `merge()` can be used to gather these files into a single data frame. In the MHP study, there were two clinicians (referred to as "junior" and "senior") who each collected a specific part of the interview. These two parts were captured in two datasets and data frames called "mhp.j" and "mhp.s." To merge these two sets into a single data frame, one must be able to track each observation so that each prisoner in "mhp.j" will be included with his data in "mhp.s." Here, each prisoner is characterized by the prison number (variable "centre") and the prisoner's "number" (each prisoner has a unique number in a given prison). The two files "mhp.j" and "mhp.s" are thus merged according to the two variables "centre" and "number":

```
> mhp <- merge(mhp.j, mhp.s, by = c("centre", "number"))
```

In a multicentre study, it may happen that datasets are captured in each centre so that data frames must be merged in blocks of observations and not in blocks of variables. The appropriate function in R to do this is `rbind()` (the "r" in `rbind()` stands for "row"). Let us suppose we have divided the mhp data frames into three subsets: prisoners younger than 40, prisoners of 40 or more, and prisoners with missing data for "age":

```
> mhp.y <- subset(mhp, age < 40)
> mhp.ny <- subset(mhp, age >= 40)
> mhp.naage <- subset(mhp, is.na(age))
```

These three data frames can be merged as follows:

```
> mhp2 <- rbind(mhp.y, mhp.ny, mhp.naage)
```

Finally, it may be useful to sort a data frame according to the values of a series of variables. For instance, if we ask for the "mhp" data frame to be sorted into increasing values of the two variables "centre" and "number" (which characterize each prisoner), then the following instructions can be used:

```
> mhp$centre
 [1]  1  1  1  1  1  1  1  1  1  1  1  1  1  1  1  1  1  1  1  1  1  1  1
[24]  1  1  1  1  1  1  1  1  1  1  1  1  1  1  1  1 10 10 10 10 10 10 10 10 10 10
[47] 10 10 11 11 11 11 11 11 11 11 11 11 11 11 11 11 11 11 11 11 11 11 11 11 11
.... [TRUNCATED]
```

```
                                                  ❶
[346] 19 19 19 19 19 19 19 19 19 19 19 19 19 19 19 19  2  2  2  2  2  2  2
[369]  2  2  2  2  2  2  2  2  2  2  2  2  2  2  2  2  2  2  2  2  2  2  2
[392]  2  2  2  2  2  2  2  2  2  2  2  2  2  2  2  2  2  2  2 20 20 20 20
.... [TRUNCATED]
> mhp$number                              ❷
  [1]  1 10 11 12 13 14 15 16 17 18  2 20 21 22 23 24 25 26 27 28 29  3 30
 [24] 31 32 33 34 35 36 37 38  4  5  6  7  8  9  1 10 11  2  3  4  5  6  7
.... [TRUNCATED]
              ❸
> x <- order(mhp$centre, mhp$number)
> x        ❹
  [1]  1 11 22  32  33  34  35  36  37   2   3   4   5   6   7   8   9
 [18] 10 12 13  14  15  16  17  18  19  20  21  23  24  25  26  27  28
 [35] 29 30 31 362 373 384 395 406 407 408 409 410 363 364 365 366 367
.... [TRUNCATED]
                       ❺
> mhp.sort <- mhp[x, ]
> mhp.sort$centre
  [1]  1  1  1  1  1  1  1  1  1  1  1  1  1  1  1  1  1  1  1  1  1  1  1
 [24]  1  1  1  1  1  1  1  1  1  1  1  1  1  1  1  2  2  2  2  2  2  2  2
 [47]  2  2  2  2  2  2  2  2  2  2  2  2  2  2  2  2  2  2  2  2  2  2  2
.... [TRUNCATED]
> mhp.sort$number
  [1]  1  2  3  4  5  6  7  8  9 10 11 12 13 14 15 16 17 18 20 21 22 23 24
 [24] 25 26 27 28 29 30 31 32 33 34 35 36 37 38  1  2  3  4  5  6  7  8  9
 [47] 10 11 12 13 14 15 16 17 18 19 20 21 22 23 24 25 26 27 28 29 30 31 32
.... [TRUNCATED]
```

Originally, the variable "centre" is not sorted ❶. The function order() ❸ is used to order the rows indexes of the data frame "mhp" in increasing values for "centre" and "number" (two prisoners in the same "centre" will be ranked according to their "number"). For instance, the 11th observation (i.e., corresponding to the 11th row of the data frame) ❹ is in second position because it corresponds to the second lowest value for "number" ❷ and to the lowest value for "centre." The whole "mhp" data frame can then be rearranged according to this ranking ❺.

9.3 Manipulation of Variables

In a few words: Changing variable names or factor labellings, turning numerical variables into factors and vice versa, grouping factor levels, or dividing continuous variables into categories—these are some operations in daily use.

In Practice: Variable names should not be changed during the statistical analyses phase, mainly for reasons of traceability of results. It is perhaps preferable to duplicate the misnamed variable to form a new one, with an appropriate name. For instance, in the mhp data frame, most diagnoses are expressed with three letters ("dep" for depression, "scz" for schizophrenia, etc.). A notable exception concerns "alcohol abuse and dependence," which is named "alcohol." If we wish to change this name into the three-letter name "alc," then the following instructions can be used:

```
> mhp$alc.mini <- mhp$alcohol.mini
```

There is now a new variable named "alc.mini" in the data frame "mhp" and it is identical to the previous variable "alcohol.mini," which is still present. This approach, however, has disadvantages. Data manipulations can be performed on "alc.mini" and, by error, "alcohol.mini" can be used instead in the analyses. In view of this, if the preference goes to changing the variable name instead of creating a new variable, this is obtained via the function names():

```
          ❶
> bckup <- mhp$alcohol.mini
> names(mhp)[names(mhp) == "alcohol.mini"] <- "alc.mini"
          ❸
> table(bckup, mhp$alc.mini, useNA = "ifany")

   0    1
0  714  0
1  0    85
```

Before making an alteration of this sort, it is a good habit to create a backup of the data frame ❶ to verify that the variable content has not changed by error. The function names() used here to convert the name "alcohol.mini" into "alc.mini." In ❸, the function table() shows that the variable content has not changed: The original variable mhp$alcohol.mini is identical to mhp$alc.mini.

Sometimes categorical variables are coded as integers during the data capture process. This is the case for the variable "profession" in the data frame mhp.s, which corresponds to variables assessed by the "senior" interviewer:

```
> str(mhp.s$prof)
int [1:799] 7 4 5 NA 6 5 4 6 7 NA ...
```

The function factor() can be used to convert this integer variable into a "factor," in other words, into a categorical variable:

```
> mhp.s$prof <- factor(mhp.s$prof)
> str(mhp.s$prof)
Factor w/ 7 levels "1","2","3","4",..: 7 4 5 NA 6 5 4 6 7 NA ...
```

The function `as.numeric()` can be used to turn a categorical variable back into its original numerical form. The syntax that must be used is, however, not straightforward:

```
> mhp.s$prof <- as.numeric(levels(mhp.s$prof))[mhp.s$prof]
> str(mhp.s$prof)
num [1:799] 7 4 5 NA 6 5 4 6 7 NA ...
```

Now considering categorical variable "prof," its levels "1," "2," ... , "7" could be set out in a more meaningful way, as "farmer," "craftsman," ..., "other." The function `levels()` is appropriate for this:

```
> bckup <- mhp.s$prof
> levels(mhp.s$prof) <- c("farmer", "craftsman", "manager",
    "intermediate", "employee", "worker", "other")
> table(bckup, mhp.s$prof, useNA = "ifany")
```

bckup	farmer	craftsman	manager	intermediate	employee	worker	other	<NA>
1	5	0	0	0	0	0	0	0
2	0	91	0	0	0	0	0	0
3	0	0	24	0	0	0	0	0
4	0	0	0	57	0	0	0	0
5	0	0	0	0	136	0	0	0
6	0	0	0	0	0	228	0	0
7	0	0	0	0	0	0	31	0
<NA>	0	0	0	0	0	0	0	227

Integer variables sometimes must be recoded. This is the case for most of the diagnoses that were originally present in the MHP database: They were coded as (yes 1/no 2) while a coding (yes 1/no 0) would be more suitable (especially because if a logistic regression is estimated via the function `glm()`, then the binary outcome variable must necessarily be coded as (0,1)). The function `ifelse()` can be used to recode a binary variable:

```
> bckup <- mhp.s$dep.cons
                              ❶                  ❷  ❸
> mhp.s$dep.cons <- ifelse(mhp.s$dep.cons == 2, 0, 1)
      ❹
> table(bckup, mhp.s$dep.cons, useNA = "ifany")

bckup    0    1
    1    0  313
    2  486    0
```

If "dep.cons" is equal to "2"❶, then the variable is recoded "0"❷; otherwise, it is "1"❸. Here again, the function `table()` is used to verify that the conversion has been performed correctly.

When the variable is not binary, the function `recode()` in the package "car" can be useful. Consider the variable "harm avoidance," as it was originally in the junior physician dataset:

```
> table(mhp.j$ha, useNA = "ifany")

   1    2    3    9  <NA>
 315  154  222    1   107
```

"1" is coded for "low," "2" for "average," "3" for "high," and "9" for "no information." There are also, as always, some missing data "NA." If we want to combine levels "1" and "2" and recode "9" as "NA," then the following instruction can be used:

```
> library(car)
                                                          ❶
> mhp.j$ha.recode <- recode(mhp.j$ha, "c(1, 2) = 1; 3 = 2;
    else = NA")
> table(mhp.j$ha.recode, mhp.j$ha, useNA = "ifany")
```

	❷		❸		
	1	2	3	9	<NA>
1	315	154	0	0	0
2	0	0	222	0	0
<NA>	0	0	0	1	107

Recoding instructions appear in a character string ❶ separated by semicolons. After a manipulation of this kind, it is highly recommended to check that the new variable `ha.recode` corresponds to what is expected. This is the objective of the call to `table()`: "1" and "2" are now grouped on the one hand and "9" and "NA" on the other.

To merely recode "9" with "NA," the function `is.na()` is the most appropriate:

```
> bckup <- mhp.j$ha
> is.na(mhp.j$ha) <- mhp.j$ha == 9
> table(bckup, mhp.j$ha, useNA = "ifany")
```

bckup	1	2	3	<NA>
1	315	0	0	0
2	0	154	0	0
3	0	0	222	0
9	0	0	0	1
<NA>	0	0	0	107

As discussed in Section 6.1, it is sometimes necessary to convert a continuous variable into a categorical variable. For example, if we wish to focus on the three populations "age 30 or under," "age between 30 and 50," and

"age strictly above 50," then the function cut() can be used to design a new
variable named "age.3c":

```
                                        ❶
> mhp.s$age.3c <- cut(mhp.s$age, breaks = c(-Inf, 30, 50, Inf),
        ❷
  labels = FALSE)
> str(mhp.s$age.3c)
❸
int [1:799] 2 2 2 2 2 2 1 2 2 2 ...
```

The instruction "breaks"❶ determines the cut-offs. The values "-Inf" and
"Inf" stand for "minus infinity" and "plus infinity"; they are used to specify
that all values smaller than 30 will correspond to the same category and the
same for all values larger than or equal to 50. The option labels = FALSE❷
provides a coding using integer values ❸ (here 1, 2, and 3). To obtain a categorical variable coded with labels "young," "med," and "old," the following
instructions would be used:

```
> mhp.s$f.age <- cut(mhp.s$age, breaks = c(-Inf, 30, 50, Inf),
  labels = c("young", "med", "old"))
> str(mhp.s$f.age)
Factor w/ 3 levels "young", "med", ..: 2 2 2 2 2 2 1 2 2 2 ...
```

9.4 Checking Inconsistencies

In a few words: In Section 9.2 we saw that impossible values (for instance,
age of over 150 years) should be checked during the univariate descriptive
part of the analysis. In this context, the two functions max() and min() are
of great value because they point immediately to the potential existence of
a problem.

Some other inconsistencies can be more difficult to apprehend. If enough
time is taken to crosscheck the responses obtained in a questionnaire study,
some very curious situations will arise, such as being born after being mar-
ried, or never having drunk any alcohol while having already been drunk.
Skip instructions are traditional sources of problems of this kind. These
skips occur, for example, when the answer to a key question implies that
the answers to the next questions are fully determined. (If the answer to
the question "Have you already drunk alcohol?" is "No," then there is no
need to look for the average alcohol consumption during the past month:
The interviewee can then go directly to the next part of the questionnaire).
Of course, some subjects will not notice the skip instruction and give an
impossible pattern of responses. This should be systematically verified.

Finally, duplicated lines are also classic sources of error found in a dataset.

In Practice: The function duplicate() is appropriate to find duplicate observations. For example, if one wants to determine if there are two identical lines in the questionnaire completed by the senior investigator in the MHP study, the following instructions can be used:

```
    ❶        ❷
> sum(duplicated(mhp.s))
[1] 0
```

Once one line of mhp.s is identical to a previous line, duplicated(mhp.s) ❷ is set to "true." The summation ❶ of all these "true" values (which are coded numerically as "1") will thus provide the number of lines in excess.

In the "mhp.s" data frame, the variable that is supposed to identify each prisoner is the "file" variable. It is thus important to verify that there are no duplicates of "file" values, in particular because the different questionnaires in the MHP study can be merged on the basis of these values:

```
> sum(duplicated(mhp.s$file))
[1] 7❶
> mhp.s$file[duplicated(mhp.s$file)]
      ❷  ❸                    ❸
[1] 185 NA 480 480 796 432 NA
```

There are seven ❶ "file" values that have several occurences. Two prisoners have "file = 185" ❷ and three have "file = NA" ❸.

There is therefore a necessity to create a new variable that will identify each prisoner. This variable can be obtained from the variable "centre" and the variable "number." In a given "centre," all prisoners have a different "number":

```
> str(mhp.s)
'data.frame   : 799 obs. of 96 variables:
$ centre      : int 1 1 1 1 1 1 1 1 1 1 ...
$ file        : int 130 122 121 142 128 127 123 125 126 129 ...
$ number      : int 1 2 3 4 5 6 7 8 9 10 ...
 .... [TRUNCATED]
> range(mhp.s$centre)
[1] 1 22❶
> range(mhp.s$number)
[1] 1 57❷
          ❸                              ❹
> mhp.s$id <- mhp.s$centre * 100 + mhp.s$number
> str(mhp.s$id)
num [1:799] 101 102 103 104 105 ...
> sum(duplicated(mhp.s$id))
[1] 0❺
```

There are 22 centres numbered from 1 to 22 ❶ and in each centre the largest value of the variable "number" is 57❷. Because this value is smaller than 100, the addition ❸ of "centre" × 100 with "number" corresponds to merging the two variables so that all prisoners should have a different value for "id" ❸. This is verified in ❺.

At this point, it may be useful to determine if any two data frames or any two variables are identical. The function identical() is dedicated to this:

```
> mhp2 <- mhp
                 ❶
> identical(mhp2, mhp)
[1] TRUE
> mhp2$age[15] <- 150
> identical(mhp2, mhp)
[1] FALSE❷
      ❸        ❹              ❺
> which(mhp != mhp2, arr.ind = TRUE)
        row    col
[1, ]   15❻   211❼
```

The use of identical() is straightforward ❶. If the two data frames are different ❷, then the function which()❸ can help determine which variables for which observation differ. The instruction "!="❹ stands for "different from." The option ❺ is necessary to obtain the index (or indices) of the variable ❼ and the observation ❻ generating the discrepancy.

Non-trivial inconsistencies can sometimes appear during close examination of the univariate description of a dataset. For instance, involving the mhp data frame:

```
> describe(mhp, num.desc = c("mean", "sd", "median", "min",
  "max", "valid.n"))
Description of mhp
```

```
Numeric
              mean      sd    median   min   max   valid.n
.... [TRUNCATED]
direct.1    0.2541   0.4356     0       0     1     799❶
direct.2    0.1715   0.3772     0       0     1     799❶
direct.3    0.1427   0.35       0       0     1     799❶
direct.4    0.3191   0.4664     0       0     1     799❶
direct.5    0.363    0.4812     0       0     1     799❶
direct      0.2059   0.4046     0       0     1     714❹
>.... [TRUNCATED]
```

The binary variable "direct" reflects a low level of Cloninger's (2000) "self-directedness," which is defined by at least three "yes" responses among the answers to "direct.1" (irresponsible), "direct.2" (purposeless), "direct.3" (helpless), "direct.4" (poor self-acceptance), and "direct.5" (poor impulse control).

Now, how is it that there are no missing data for "direct.1," "direct.2," ..., "direct.5" (the sample size of the MHP study is 799 ❶), while there are 85 missing data for "direct" ❷? To solve this apparent paradox, we had to get back to the company that captured the data to discuss the issue. This revealed that when an interview had to stop prematurely due to local constraints, the end of the questionnaire was abandoned and the interviewers, using a missing data key, only answered the global question "direct" and not each of the items "direct.1," "direct.2," etc. When these answers were captured, they were imputed by error as "0" and not as "NA." If, for a given prisoner, there is missing data for "direct," then we need to replace the corresponding values of "direct.1," "direct.2," ..., "direct.5" by the missing data key "NA":

```
> is.na(mhp$direct.1) <- is.na(mhp$direct)
> is.na(mhp$direct.2) <- is.na(mhp$direct)
> is.na(mhp$direct.3) <- is.na(mhp$direct)
> is.na(mhp$direct.4) <- is.na(mhp$direct)
> is.na(mhp$direct.5) <- is.na(mhp$direct)
> names.direct <- c("direct.1", "direct.2", "direct.3",
    "direct.4", "direct.5")
> describe(mhp[, names.direct], num.desc = c("mean", "sd",
    "median", "min", "max", "valid.n"))
Description of mhp[, names.direct]
```

```
Numeric
            mean      sd     median   min   max   valid.n
direct.1   0.2773   0.448       0      0     1     714❶
direct.2   0.1863   0.3896      0      0     1     714❶
direct.3   0.1541   0.3613      0      0     1     714❶
direct.4   0.3557   0.4791      0      0     1     714❶
direct.5   0.3922   0.4886      0      0     1     714
```

As expected, all six variables now have the same amount of missing data ❶.

In the "junior" questionnaire, there are two questions concerning a possible history of past imprisonment. A binary question (past.prison) enquires if past imprisonment has occurred or not and, if the response is "yes," then another question (n.prison) determines how many times the prisoner has been in prison. In theory, if "n.prison" is 1 or more, then "past.prison" should not be "no" (coded here as a "0"). In practice,

```
> table(mhp$past.prison, mhp$n.prison, useNA = "ifany")
```

	1	2	3	4	5	6	7	8	9	10	11	12	13	14	15	16
0	0	20❶	20❶	0	0	10❶	0	0	0	0	0	0	0	0	10❶	0
1	138	69	37	32	25	13	8	4	5	19	1	1	3	1	7	1
<NA>	0	0	0	0	0	0	0	0	0	0	0	0	0	0	0	0

```
17     20 21   24   25 <NA>
 0      0  0    0    0    0  403
 1      1  3    2    1    1   14
<NA>    0  0    0    0    0    4
```

Obviously, all results corresponding to ❶ are inconsistencies. They need to be solved, and it is decided that if "n.prison" is strictly above 0, then "past.prison" is necessarily "yes" (coded as "1"):

```
> mhp$past.prison[mhp$n.prison > 0] <- 1
> table(mhp$past.prison, mhp$n.prison, useNA = "ifany")

       1   2   3   4   5   6   7   8   9  10  11  12  13  14  15  16
0      0   0   0   0   0   0   0   0   0   0   0   0   0   0   0   0
1    138  71  39  32  25  14   8   4   5  19   1   1   3   1   8   1
<NA>   0   0   0   0   0   0   0   0   0   0   0   0   0   0   0   0

      17 20 21 24 25 <NA>
0      0  0  0  0  0  403
1      1  3  2  1  1   14
<NA>   0  0  0  0  0    4
```

In the "junior" questionnaire, there is a structured interview called the "M.I.N.I." (Mini-International Neuropsychiatric Interview) (Sheehan et al. 1995). This instrument operationalizes psychiatric diagnosis in a very formal manner. For instance, for depression, two main symptoms are considered first: "sadness" and "anhedonia" (defined as an inability to experience pleasure from normally pleasurable life events). If none of the questions relating to these two symptoms obtains a positive response, the interviewer can skip the section and go to the following mental disorder. Otherwise, the interviewer must ask seven other binary questions; and if at least four of these new symptoms are present, then the prisoner is considered to present a major depressive episode (according the M.I.N.I.). Because the objective of the MHP study is essentially to determine the prevalence of mental disorders in French prisons for men, the consistency of the algorithms used in the M.I.N.I. must be explored. First, let us have a look at the skip that occurs after the first two questions about sadness (dsm.dep1) and anhedonia (dsm.dep2):

                         ❶   ❷

```
> table(mhp$dsm.dep1 | mhp$dsm.dep2, mhp$dsm.dep12,
    useNA = "ifany", deparse.level = 2)
mhp$dsm.dep12
```

```
mhp$dsm.dep1 | mhp$dsm.dep2      0        1     <NA>
                       FALSE    341        0     42❹
                        TRUE     6❸     353     55❹
                       <NA>       2        0      0
```

The objective is to determine if prisoners who present at least one of these two symptoms (the sign "|" in ❶ corresponds to the logical operator "or") are also those for whom the full range of depressive symptoms is explored (this corresponds to a "1" for variable "dsm.dep12"❷). There are obviously several problems❸, ❹: six❸ prisoners have discordant values of ❶ and ❷, and has as many as 97 missing values❹ while the other has none. Do these problems concern only the intermediate variable "dsm.dep12," or do they also affect the diagnosis of depression itself (which corresponds to the variable "dep.mini")?

```
              ❶
> dep1to9 <- c("dsm.dep1", "dsm.dep2", "dsm.dep3", "dsm.dep4",
    "dsm.dep5", "dsm.dep6", "dsm.dep7", "dsm.dep8", "dsm.dep9")
                      ❷                     ❸      ❹
> dsm.dep1to9 <- apply(mhp[, dep1to9], 1, sum, na.rm = TRUE)
              ❺                              ❻
> table((mhp$dsm.dep1 | mhp$dsm.dep2) & (dsm.dep1to9 > 4),
    mhp$dep.mini, useNA = "ifany", deparse.level = 2)
                                               mhp$dep.mini
(mhp$dsm.dep1 | mhp$dsm.dep2) & (dsm.dep1to9 > 4)    0       1
                                     FALSE  505      0❼
                                      TRUE    0❼   294
```

The new variable "dep1to9"❶ gathers all the variable names of the nine items corresponding to the depression items in the M.I.N.I. The function apply()❶ is used here to sum❹ these nine items (❸ stipulates that rows are summed and not columns). The corresponding variable called "des.dep1to9" is then used in the algorithm that determines if a prisoner has a depressive disorder or not: At least one of the two items 1 and 2 must be rated "1"❺ (1 corresponds to the presence of the corresponding symptom), and at least five items among the nine depression items must also be rated "1"❻. Fortunately, it is noted that the result of this algorithm leads to results that are identical to those for the variable "dep.mini"❼.

Appendix: The Analysis of Questionnaire Data using R: *Memory Card*

A.1 Data Manipulations

A.1.1 Importation/Exportation of Datasets

```
read.csv("c:/.../toto.csv")          # import file in text format
read.csv2("c:/.../toto.csv")         # csv: coma separated value
read.delim("c:/.../toto.tab")        # delim: tab separated value
read.delim2("c:/.../toto.tab")       # csv2 or delim2 for specific
                                       countries
odbc.toto <- odbcConnectExcel("c:/.../toto.xls")       # library(RODBC)
sqlTables(odbc.toto)$TABLE_NAME                         # gives nameoftable
totoxls <- sqlFetch(odbc.toto, sqtable = "nameoftable") # import Excel file
read.spss("c:/.../toto.sav")         # read spss file
read.dta("c:/.../toto.dta")          # read stata file
write.csv(toto, "c:/.../toto.csv")   # write file in csv format
write.csv2(toto, "c:/.../toto.csv")  # same for countries where decimals
                                       are ","
save(toto1, toto2, "c:/.../toto")    # save a series of data frame
load("c:/.../toto")                  # read an object saved by save()
str(toto)                            # informations concerning an R oject
```

A.1.2 Manipulation of Datasets

```
toto.young <- toto[toto$age < 18, ]  # selection of observations
toto.somevar <- toto[, c("namevar1", "namevar2", "namevar3")]
                                       # selection of variables
toto2 <- subset(toto, age < 18, select = c("namevar1", "namevar2")]
                                       # other function for selection
toto <- merge(totovar1, totovar2, by = c("id1", "id2"))
                                       # merge datasets (add variables)
toto <- rbind(totoobs1, totoobs2)    # merge datasets (add observations)
toto.sort <- toto[order(toto$id), ]  # sort dataset according to "id"
sum(duplicated(toto))                # number of duplicated obervations
toto[duplicated(toto)]               # which obervations are duplicated
identical(toto, toto2)               # are two dataframes the same
which(toto != toto2, arr.ind = TRUE) # differences between two data frames
```

A.1.3 Manipulation of Variables

```
toto$var1.fact <- factor(toto$var1)  # transforms as a categorical variable
toto$var1 <- as.numeric(levels(toto$var1.fact))[toto$var1.fact]
                                       # backtransformation as a number
```

```
levels(toto$var1.fact) <- c("lev1", "lev2", ...)
                                      # change levels
toto$var1bin <- ifelse(toto$var1 < 20, 1, 0)
                                      # recode into a binary variable
toto$var1recode <- recode(toto$var1, "cond1; cond2; ...")
                                      # recode a variable; library(car)
is.na(toto$var1) <- toto$var1 == 9   # transform 9 a mining data
toto$var1cut <- cut(toto$var1, breaks = c(-Inf, lim1, ..., limq, Inf),
   labels = FALSE)                    # cut a numerical variable into pieces
   [contrasts() and relevel() are in the section statistical modelling]
```

A.2 Descriptive Statistics

A.2.1 Univariate

```
summary(toto)                         # mean, median, minimum, etc.
describe(toto) #library(prettyR)      # mean, median, sd, etc.
tab <- table(toto$var1, toto$var2, deparse.level = 2, useNA = "ifany")
                                      # crosstabulation of 2 variables
prop.table(tab, 1)                    # crosstabulation with %
by(toto$varcont1, toto$varcat1, mean, na.rm = TRUE)
                                      # subgroup analysis (mean)
hist(toto$var1)                       # histogram
plot(density(toto$varcont1, na.rm = TRUE))        # density curve
int <- hist(toto$varcont1, freq = FALSE, plot = FALSE)
hist(toto$varcont1, xlim = range(c(dest$x, int$breaks)),
   ylim = range(c(dest$y, int$density)), freq = FALSE)
   lines(dest, lty = 2, lwd = 2)
box()                                 # histogram and density curve
qqnorm(toto$varcont1); qqline(toto$varcont1)
                                      # normal probability plot
barplot(table(toto$varcat1))          # barplot
boxplot(toto$varcont1 ~ toto$var1cat) # boxplots in subgroups
plot(toto$varcont1 ~ jitter(toto$var1cat))
                                      # distribution in subgroups
plotmeans(toto$varcont1 ~ toto$time)  # temperature diagram; library(gplots)
```

A.2.2 Bivariate

```
twoby2(toto$exposure,tot$outcome)     # odds-ratios and RR; library(Epi)
cor(toto[, c("var1", ..., "varp")], use = "complete.obs")
                                      # correlation matrix
rcor.test(toto[, c("var1", ..., "varp")], use = "pairwise.complete.obs")
                                      # correlation matrix; library(ltm)
plot(toto$var1, toto$var2)            # cartesian diagram
plot(jitter(toto$var1), jitter(toto$var2))
abline(lm(toto$var2 ~ toto$var1, data = mhp), lwd = 2, lty = 2)
nona <- !(is.na(toto$var1) | is.na(toto$var2))
lines(lowess(toto$var1[nona], toto$var2[nona]), lwd = 2)
                                      # cartesian diagram with regression
```

A.2.3 Multidimensional

```
cha <- hclust(dist(t(scale(toto[, c("var1", ..., "varp")])))), method = "ward")
plot(cha)                          # hierarchical clustering
obj <- cor(toto[, c("var1", ..., "varp")], use = "pairwise.complete.obs")
heatmap(obj)                       # shaded representation of correlations
mdspca(toto[, c("var1", ..., "varp")]) # PCA representation of a correlation
                                   # matrix; library(psy)
sphpca(toto[, c("var1", ..., "varp")]) # spherical representation of a
                                   # correlation matrix; library(psy)
fpca(outcome ~ exposure1 + ... + exposure, data = toto)
                                   # focused PCA; library(psy)
```

A.3 Statistical Inference

```
y <- na.omit(toto$varbin1)
binom.confint(x = sum(y), n = length(y), method = "Wilson")
                                   # CI of a proportion; library(binom)
chisq.test(toto$varcat1, toto$varcat2, correct = FALSE)
                                   # comparison of two proportions
fisher.test(toto$varcat1, toto$varcat2)
                                   # comparison of two proportions
t.test(toto$varcont1 ~ toto$varbin1, var.equal = TRUE)
                                   # comparison of two means
wilcox.test(toto$varcont1 ~ toto$varbin1)
                                   # comparison of two means
cor.test(toto$var1, toto$var2)     # test that a correlation is zero
events <- table(toto$varbin1, toto$varord1)[, 2]
trials <- events + table(toto$varbin1, toto$varord1)[, 1]
prop.trend.test(events,trials)     # chisquare test for trend
n.for.survey(p = 0.01, delta = 0.02)  # sample size for a survey
n.for.2p(p1 = 0.125, p2 = 0.05, power = 0.9, ratio = 3)
                                   # sample size for a survey
```

A.4 Statistical Modelling

```
mod <- lm(outcomecont ~ var1 + ... + varp, data = toto)
summary(mod)                       # linear regression, ANOVA
drop1(mod, .~., test = "F")        # "type III" hypothesis testing
plot(mod,2)                        # regression diagnosis
plot(mod,1)                        # regression diagnosis
plot(mod,4)                        # regression diagnosis
mod <- glm(outcome01 ~ var1 + ... + varp, data = toto, family = "binomial")
summary(mod)                       # logistic regression
exp(coefficients(mod))             # adjusted odd-ratios
exp(confint(mod))                  # CI of adjusted OR
drop1(mod, .~., test = "Chisq")    # "type III" hypothesis testing
plot(mod,4)                        # regression diagnosis
```

```
logistic.display(mod, decimal = 3)    # display log reg; library(epicalc)
summary(multinom(outcomecat ~ var1 + ... + varp, data = toto))
                                      # multinomial log reg; library(nnet)
summary(polr(outcomeord ~ var1 + ... + varp, data = toto))
                                      # polytomic log reg; library(MASS)
summary(glm(outcomecount ~ var1 + ... + varp, data = toto,
   family = "quasipoisson")
                                      # Poisson regression (overdispersion)
summary(glm.nb(outcomecount ~ var1 + ... + varp, data = toto)
                                      # negative binomial reg; library(MASS)
plot(gam(outcomecont ~ s(varcont1), data = toto))
                                      # regression spline; library(mgcv)
plot(gam(outcome01 ~ s(varcont1), data = toto, family = binomial))
                                      # regression spline; library(mgcv)
summary(glm(outcome01 ~ poly(varcont1, degree = 3),data = toto,
   family = binomial))
                                      # orthogonal polynomials
contrasts(toto$varcat1)               # coding of a factor
contrasts(toto$varcat1) <- contr.sum
                                      # 1 -1 coding
contrasts(toto$varcat1) <- contr.treatment
                                      # 0 1 coding
toto$varcat1 <- relevel(toto$varcat1, ref = "newreflevel")
                                      # change level of reference of a factor
estimable(mod, c(0, ..., 0, 1, ..., -1, ..., 0))
                                      # estimation and test of a contrast
summary(step(mod, scope = list(upper = ~var1 + ... + varp,
   lower = ~varb1 + ... + varbq), trace = FALSE))
                                      # stewise regression
scale(toto)                           # normalize data set
calc.relimp(mod, type = c("lmg", "last"), rela = TRUE)
                                      # relative importance of predictors
plot(naclus(toto))                    # pattern of miss data; library(Hmisc)
micetoto <- mice(toto, seed = 1)      # impute 5 date sets; library(mice)
imputetoto <- complete(micetoto)      # completed dataset; library(mice)
summary(glm.mids(outcome01 ~ var1, ..., varp, data = micetoto,
   family = "binomial")              # multiple imputations, library(mice)
lm.boot <- function(data, index) {
mhp.boot <- data[index, ]
mod <- lm(outcomecont ~ var1 + ... + varp, data = mhp.boot)
coefficients(mod)
}
resboot <- boot(mhp.mod, lm.boot, 10000)
boot.ci(resboot, index = 2, type = "bca")
                                      # bootstrapped coefs; library(boot)
summary(glmer(outcome01 ~ var1 + ... + varp + (1 | varrandomeffect),
   data = toto, family = "binomial")  # random effect; library(lme4)
```

A.5 Validation of a Composite Score

```
toto$sumvar1p <- apply(toto[, c("var1", ..., "varp")], 1, sum, na.rm = TRUE)
                                      # summation of p items
```

```
mtmm(toto, list(c("var11", ..., "var1p"), ..., c("varq1", ..., "varqp"),
    itemTot = TRUE, namesDim = c("namescale1", ..., "namescaleq"))
                                        # multitraits multimethods;
                                        # library(psy)
scree.plot(toto[, c("var1", ..., "varp")], simu = 20)
                                        # scree plot; library(psy)
resfa <- factanal(na.omit(toto[, c("var1", ..., "varp")], factors = k)
print(resfa, cutoff = 0)                # factor analysis
print.psych(promax(loadings(resfa)), cut = 0)
                                        # oblique rotation; library(psych)
cronbach(toto[, c("var1", ..., "varp")])
                                        # Cronbach alpha; library(psy)
ckappa(toto[, c("rater1", "rater2")])   # Cohen kappa; library(psy)
lkappa(toto[, c("rater1", ..., "raterq")])
                                        # kappa for q raters; library(psy)
wkappa(toto[, c("rater1", "rater2")])   # weighted kappa; library(psy)
icc(toto[, c("rater1", "rater2")])      # intraclass correlation; library(psy)
```

References

Agresti, A. and B. A. Coull (1989). Approximate is better than "exact" for interval estimation of binomial proportions. *The American Statistician*, 52(2): 119–126.

Balakrishnan, N. and A. Childs (2001). Outlier. *Encyclopaedia of Mathematics*. M. Hazewinkel, Ed. Boston, MA: Kluwer Academic Publishers.

Berk, R. and J. MacDonald (2008). Overdispersion and Poisson regression. *Journal of Quantitative Criminology*, 24(3): 269–284.

Box, G. E. P. and N. R. Draper (1987). *Empirical model-building and response surfaces*. New York: Wiley.

Brown, L. D., T.T. Cai, and A. DasGupta (2001). Interval estimation for a binomial proportion. *Statistical Science*, 16(2): 101–133.

Campbell, D. and D. Fiske (1959). Convergent and discriminant validation by the multitrait-multimethod matrix. *Psychological Bulletin*, 56(2): 81–105.

Cattel, R. B. (1966). The scree test for the number of factors. *Multivariate Behavioral Research*, 1: 245–276.

Cloninger, C. R. (2000). A practical way to diagnosis personality disorder: A proposal. *Journal of Personality Disorders*, 14(2): 99–108.

Cohen, J. (1960). A coefficient of agreement for nominal scales. *Educational and Psychological Measurement*, 20(1): 37–46.

Cohen, J. (1988). *Statistical power analysis for the behavioral sciences* (2nd ed.). Hillsdale, NJ: Erlbaum.

Cohen, J. and P. Cohen (1975). *Applied multiple regression/correlation analysis for the behavioural sciences*. New York: Lawrence Erlbaum Associates.

Conger, A. J. (1980). Integration and generalization of kappas for multiple raters. *Psychological Bulletin*, 88: 322–328.

Cox, D. and N. Wermuth (1992). A comment on the coefficient of determination for binary responses. *The American Statistician*, 46(1): 1–4.

Cox, D. R. and N. Wermuth (1996). *Multivariate dependencies: Models, analysis and interpretation*. Boca Raton, FL: CRC Press.

Cribari-Neto, F. and S. G. Zarkos (1999). Bootstrap methods for heteroskedastic regression models: Evidence on estimation and testing. *Econometric Reviews*, 18(211): 211–228.

Cronbach, L. J. (1951). Coefficient alpha and the internal structure of tests. *Psychometrika*, 6: 297–334.

D'Agostino, R., W. Chase, and A. Belanger (1988). The appropriateness of some common procedures for testing the equality of two independent binomial populations. *The American Statistician*, 42(3): 198–202.

Dalton, M. A., J. D. Sargent, M.L. Beach, L. Titus-Ernstoff, J. Gibson, M.B. Ahrens, J. J. Tickle, and T. F. Heatherton (2003). Effect of viewing smoking in movies on adolescent smoking initiation: A cohort study. *Lancet*, 362(9380): 281–285.

Davis, L. J. (1985). Consistency and asymptotic normality of the minimum logit chi-squared estimator when the number of design points is large. *The Annals of Statistics*, 13(3): 947–957.

Efron, B. and R. Tibshirani (1993). *An introduction to the bootstrap*. Boca Raton, FL: CRC Press.

Falissard, B. (1996). A spherical representation of a correlation matrix. *Journal of Classification*, 13(2): 276–280.

Falissard, B., J.-Y. Loze, I. Gasquet, A. Duburc, C. De Beaurepaire, F. Fagnani, and F. Rouillon (2006). Prevalence of mental disorders in French prisons for men. *BMC Psychiatry*, 6: 33.

Faraway, J. J. (2006). *Extending the linear model with R*. Boca Raton, FL: Chapman & Hall/CRC.

Flanders, W. D. and K. J. Rothman (1982). Interaction of alcohol and tobacco in laryngeal cancer. *American Journal of Epidemiology*, 115(3): 371–379.

Fowler, F. J. (2008). *Survey research methods (applied social research methods)*. Thousand Oaks, CA: Sage Publications.

Fox, J. (2002). An R and S-plus companion to applied regression. Thousand Oaks, CA: Sage Publications.

Gardner, W., E. P. Mulvey, and E. C. Shaw (1995). Regression analyses of counts and rates: Poisson, overdispersed Poisson, and negative binomial models. *Psychological Bulletin*, 118(3): 292–304.

Gelman, A. and J. Hill (2008). Data analysis using regression and multilevel/hierarchical models. Cambridge, UK: Cambridge University Press.

Ghosh, S., D. Pahwa, and D. C. Rennie (2008). Comparison of design-based and model-based methods to estimate the variance using National Population Health Survey data. *Model Assisted Statistics and Applications*, 3(1): 33–42.

Goodman, S. N. (1993). *p* values, hypothesis tests, and likelihood: Implications for epidemiology of a neglected historical debate. *American Journal of Epidemiology*, 137(5): 485–496.

Graham, P. and J. Jackson (1993). The analysis of ordinal agreement: beyond weighted kappa. *Journal of Clinical Epidemiology*, 46: 1055–1062.

Graubard, B. I. and E. L. Korn (1996). Modelling the sampling design in the analysis of health surveys. *Statistical Methods in Medical Research*, 5(3): 263–281.

Greenacre, M. and J. Blasius (2006). *Multiple correspondence analysis and related methods*. London: Chapman & Hall/CRC.

Greenland, S., M. Maclure, J. J. Schlesselman, C. Poole, and H. Morgenstern (1991). Standardized regression coefficients: A further critique and review of some alternatives. *Epidemiology*, 2(5): 387–392.

Greenland, S., J. J. Schlesselman, and M. H. Criqui (1986). The fallacy of employing standardized regression coefficients and correlations as measures of effect. *American Journal of Epidemiology*, 123(2): 203–208.

Grömping, U. (2006). Relative importance for linear regression in R: The package relaimpo. *Journal of Statistical Software*, 17(1).

Grömping, U. (2007). Estimators of relative importance in linear regression based on variance decomposition. *The American Statistician*, 61(2): 139–147.

Harrel, F. E. (2001). *Regression modelling strategies with applications to linear models, logistic regression, and survival analysis*. New York: Springer.

Hill, A. B. (1965). The environment and disease: Association or causation? *Proceedings of the Royal Society of Medicine*, 58: 295–300.

Hills, M. (1969). On looking at large correlation matrices. *Biometrika*, 56(2): 249–253.

Horn, J. L. (1965). A rationale and test for the number of factors in factor analysis, *Psychometrika*, 30: 179–185.

Horton, N. J. and S. R. Lipsitz (2001). Multiple imputation in practice: Comparison of software packages for regression models with missing variables. *American Statistician*, 55(3): 244–254.

Hosmer, D. W., T. Hosmer, S. Le Cessie, and S. Lemeshow, S. (1997). A comparison of goodness-of-fit tests for the logistic regression model. *Statistics in Medicine*, 16: 965–980.

Hosmer, D. W. and S. Lemeshow (1989). *Applied logistic regression*. New York: John Wiley & Sons.

Hu, L. T. and P. M. Bentler (1999). Cutoff criteria for fit indexes in covariance structure analysis: Conventional criteria versus new alternatives. *Structural Equation Modelling*, 6(1): 1–55.

Jackson, J. E. (1991). *A user's guide to principal components*. New York: Wiley.

Jones, M. P. (1996). Indicator and stratification methods for missing explanatory variables in multiple linear regression. *Journal of the American Statistical Association*, 91(433): 222–230.

Koopman, R. (2008) Interpreting main effects in regression models with interactions. http://sci.tech–archive.net/Archive/sci.stat.math/2008–05/msg00264.html.

Lance, C. E., M. M. Butts, and L. C. Michels (2006). The sources of four commonly reported cutoff criteria. What did they really say? *Organizational Research Methods*, 9(2): 202–220.

Landis, J. R. and G. G. Koch (1977). The measurement of observer agreement for categorical data. *Biometrics*, 33: 159–174.

Lebart, M., A. Morineau, and M. Piron (1995). *Statistique exploratoire multidimensionnelle*. Paris: Dunod.

Lehman, E. L. (1993). The Fisher, Neyman-Pearson theories of testing hypotheses: One theory or two? *Journal of the American Statistical Association*, 88(424): 1242–1249.

Lemeshow, S., L. Letenneur, J.-F. Dartigues, S. Lafont, J.-M. Orgogozo, and D. Commenges (1998). Illustration of analysis taking into account complex survey considerations: The association between wine consumption and dementia in the PAQUID study. *American Journal of Epidemiology*, 148(3): 298–306.

Loehlin, J. C. (1987). *Latent variable models: An introduction to factor, path, and structural analysis*. Hillsdale, NJ: Erlbaum.

MacCallum, R. C., M. W. Browne, and H. M. Sugawara (1996). Power analysis and determination of sample size for covariance structure modelling. *Psychological Methods*, 1(2): 130–149.

Maindonald, J. and J. Braun (2007). *Data analysis and graphics using R: An example-based approach*. Cambridge, UK: Cambridge University Press.

McDonald, R. P. (1981). The dimensionality of tests and items. *British Journal of Mathematical and Statistical Psychology*, 34: 100–117.

McNemar, Q. (1946). Opinion-attitude methodology. *Psychological Bulletin*, 43: 289–374.

Menard, S. (2004). Six approaches to calculating standardized logistic regression coefficients. *The American Statistician*, 58: 218–223.

Mood, C. (2009). Logistic regression: Why we cannot do what we think we can do, and what we can do about it. *European Sociological Review Advanced Access*, doi: 10.1093/esr/jcp006. *European Sociological Review*, 26(1): 67–82, 2010.

Nunnally, J. C. and I. H. Bernstein (1994). *Psychometric theory*. New York: McGraw-Hill.

O'Grady, K. E. (1982). Measures of explained variance: Cautions and limitations. *Psychological Bulletin*, 92(3): 766–777.

Ozer, D. J. (1985). Correlation and the coefficient of determination. *Psychological Bulletin*, 97(2): 307–315.

Peduzzi, P., J. Concato, E. Kemper, T. R. Holford, and A. R. Feinstein (1996). A simulation study of the number of events per variable in logistic regression analysis. *Journal of Clinical Epidemiology*, 49(12): 1373–1379.

Raftery, A. E. (1993). Bayesian model selection in structural equation models. In *Testing structural equation models*. K. A. Bollen and J. S. Long, Editors. Newbury Park, CA: Sage: pp. 163–180.

Roberts, J. K. and X. Fan (2004). Bootstrapping within the multilevel/hierarchical linear modelling framework: A primer for use with SAS and SPLUS. *Multiple Linear Regression Viewpoints*, 30(1): 23–34.

Rothman, K. J. and S. Greenland (1998). *Modern epidemiology*. Philadelphia, PA: Lippincott, Williams & Wilkins.

Rowe, A. K., M. Lama, F. Onikpo, and M. S. Deming (2002). Design effects and intraclass correlation coefficients from a health facility cluster survey in Benin. *International Journal of Quality Health Care*, 14(6): 521–523.

Rutherford, A. (2001). *Introducing ANOVA and ANCOVA: A GLM approach*. Thousand Oaks, CA: Sage.

Schwartz, D. (1986). *Méthodes statistiques à l'usage des médecins et des biologistes*. Paris, Flammarion Médecine Sciences, p. 318.

Seigel, D. G., M. J. Podgor, and N.A. Remaley (1992). Acceptable values of kappa for comparison of two groups. *American Journal of Epidemiology*, 135: 571–578.

Sheehan, D. V., Y. Lecrubier, K. H. Sheehan, P. Amorim, J. Janavs, E. Weiller, T. Hergueta, R. Baker, and G.C. Dunbar (1998). The Mini-International Neuropsychiatric Interview (M.I.N.I.): The development and validation of a structured diagnostic psychiatric interview for DSM-IV and ICD-10. *Journal of Clinical Psychiatry*, 59(Suppl. 20): 22–33.

Shrout, P. E. and J. L. Fleiss (1979). Intraclass correlations: uses in assessing rater reliability. *Psychological Bulletin*, 86: 420–428.

Smith, G. A. and G. Stanley (1983). Clocking g: Relating intelligence and measures of timed performance. *Intelligence*, 7: 353–368.

Snedecor, G. W. and W. G. Cochran (1989). *Statistical methods*. Ames, IA: Iowa State University Press.

Snijders, T. A. B. and R. J. Bosker (1999). *Multilevel analysis: An introduction to basic and advanced multilevel modelling*. Thousand Oaks, CA: Sage.

Spence, I. (2005). No humble pie: The origins and usage of a statistical chart. *Journal of Educational and Behavioral Statistics*, 30(4): 353–368.

Strauss, R. S. (1999). Comparison of measured and self-reported weight and height in a cross-sectional sample of young adolescents. *International Journal Obesity and Related Metabolic Disorders*, 23: 904–908.

Tacq, J. (1997). Multivariate analysis techniques in social science research: From problem to analysis. Thousand Oaks, CA: Sage.

Thurston, D. L. (2006). Multi-attribute utility analysis of conflicting preferences. In *Decision making in engineering design*, E. L. Kemper, W. Chen, and L. C. Schmidt, Editors. New York: ASME Press.

Uebersax, J. S. (1987). Diversity of decision-making models and the measurement of interrater agreement. *Psychological Bulletin*, 101: 140–146.

Ukoumunne, O. C., M. C. Gulliford, S. Chinn, J. A. C. Sterne, and P. G. J. Burney (1999). Methods for evaluating area-wide and organisation-based interventions in health and health care: A systematic review. *Health Technology Assessment,* 3(5): iii–92.

Van Buuren, S., J. P. L. Brand, C. G. M. Groothuis-Oudshoorn, and D.B. Rubin (2006). Fully conditional specification in multivariate imputation. *Journal of Statistical Computation and Simulation,* 76(12): 1049–1064.

Venables, W. N. and B. D. Ripley (2002). *Modern applied statistics with S.* Berlin: Springer.

Ver Hoef, J. M. and P. L. Boveng (2007). Quasi-Poisson vs. negative binomial regression: How should we model overdispersed count data? *Ecology,* 88(11): 2766–2772.

Vittinghoff, E. and C. E. McCulloch (2007). Relaxing the rule of ten events per variable in logistic and Cox regression. *American Journal of Epidemiology,* 165(6): 710–718.

Wikipedia contributors (2009a). "Heteroscedasticity."

Wikipedia contributors (2009b). "Poisson distribution."

Wright, S. (1921). Correlation and causation. *J. Agricultural Research,* 20: 557–585.

Young, G. A. (1994). Bootstrap: More than a stab in the dark? *Statistical Science,* 9: 382–418.

Zeileis, A., C. Kleiber, and S. Jackman. (2008). Regression models for count data in R. *Journal of Statistical Software,* 27(8): 1–25.

Zwick, R. (1988). Another look at interrater agreement. *Psychological Bulletin,* 103: 374–378.

Index